The Urban Book Series

The Urban Book Series is a resource for urban studies and geography research worldwide. It provides a unique and innovative resource for the latest developments in the field, nurturing a comprehensive and encompassing publication venue for urban studies, urban geography, planning and regional development.

The series publishes peer-reviewed volumes related to urbanization, sustainability, urban environments, sustainable urbanism, governance, globalization, urban and sustainable development, spatial and area studies, urban management, transport systems, urban infrastructure, urban dynamics, green cities and urban landscapes. It also invites research which documents urbanization processes and urban dynamics on a national, regional and local level, welcoming case studies, as well as comparative and applied research.

The series will appeal to urbanists, geographers, planners, engineers, architects, policy makers, and to all of those interested in a wide-ranging overview of contemporary urban studies and innovations in the field. It accepts monographs, edited volumes and textbooks.

Now Indexed by Scopus!

More information about this series at http://www.springer.com/series/14773

Luciano Crespi
Editor

Design of the Unfinished

A New Way of Designing Leftovers Regeneration

Springer

Editor
Luciano Crespi
Dipartimento di Design
Politecnico di Milano
Milan, Italy

ISSN 2365-757X ISSN 2365-7588 (electronic)
The Urban Book Series
ISBN 978-3-030-73456-5 ISBN 978-3-030-73457-2 (eBook)
https://doi.org/10.1007/978-3-030-73457-2

This Springer imprint is published by the registered company Springer Nature Switzerland AG
The registered company address is: Gewerbestrasse 11, 6330 Cham, Switzerland

Introduction

For almost 40 years, the reuse of large industrial areas that are no longer active has been dealt with all over the world. Italy was among the first countries to have a very influential school of thought in the field of urban transformations induced by the disposal of entire industrial sectors and in the study of how to implement their rehabilitation (Bianchetti 1985; Secchi et al. 1985). Next came the reflections, and above all the policies, from other industrialised countries affected by the same phenomenon (Chaix 1989; Roberts and Sykes 2000). The literature about the re-functionalisation of former industrial areas or other types of disused spaces is now considerable, but in its vast majority, it concerns examples of building renovation. More recently, studies have also been published about interventions based on participatory procedures, therefore involving as protagonists a plurality of subjects often belonging to new creative 'classes'. The nature of these publications is mainly descriptive. The method adopted is that of the 'case study'. On this subject, what is missing is a design thinking which goes beyond the observation of opportunity reasons—ethical, economic, cultural and social—that guided the decision of recovering these places and memory immense heritage, or the effectiveness of these interventions also in terms of public and critical consensus. A design thinking capable of indicating a perspective and some work trajectories, demonstrating the fascination that these architectures still preserve.

In the panorama of studies on this subject, *Design of the unfinished* represents a distinctive survey on the regeneration of *leftovers*—warehouses, churches, schools, stations, etc...—i.e. places that have stopped playing the role for which they were built. The experimental design work carried out for many years with interior design students at Politecnico di Milano, has made it possible to formulate a paradigm which can offer an alternative to both the practices of building renovation adopted up to now and the experiments of temporary use carried out in the form of self-design and often even self-construction. While in the old paradigm can always be found the clues necessary to overcome it, however in the history of knowledge there are passages when a leap is required, a programmatic break with past knowledge. This is what these chapters propose. Evidence coming from the design universe is given that testifies the diffusion, all over the world, of first signs of a design approach to the of leftovers regeneration that belongs to what is defined here as 'design of the unfinished'.

This method of intervention is also aimed at countering the gentrification phenomenon, which Giovanni Semi has addressed in an important text of great scientific accuracy (Semi 2015). Interventions promoted by land speculation, designed to subtract parts of the historic nuclei from the original residents' use and assign them to a new cultural class searching for glamorous places, are opposed by an approach capable of enhancing the *leftovers* memory stratification, tasking them to respond to the hospitality and relevant services demand from the new border crossers, the contemporary neo-nomads. It is a matter of operating capillary, not replacing entire urban areas, but through minimally invasive surgery, to give them new temporary functions, capable of adapting to the needs arising from emergencies. As the Pandemic has taught us in recent months, most of the mega-urban regeneration projects, destined to take up mountains of economic resources and require biblical timeframes, risk becoming obsolete before they are even born. Faced with a world that is changing at an astonishing speed and seems unmanageable with the tools we have known up to now, it looks wise adopting in cities measures that respond to a strategic design but can be implemented in a short time, with a minimum of resources, are reversible and do not involve further land consumption. At the same time, they should also witness the need for a new aesthetic, not only in the arts sphere, as Nicolas Bourriaud loudly calls for in his recent *Inclusions. Aesthetics of the Capitalocene* (Bourriaud 2020), but also in design and architecture.

To strengthen the hypotheses put forward by the working group that in recent years has carried out the didactic experimentation at the Politecnico di Milano, testimonies coming from other geographical and disciplinary contexts are collected in the second part of the book.

Chapter 1, *Manifesto of design of the unfinished*, identifies neo-nomadism as an issue characterising this beginning of the century and as a challenging field for the project culture. It calls for questioning the meaning of living in the contemporary world, in post-industrial cities, in relation to available spaces no longer used, which are named leftovers. Their regeneration project is described, which can assume their degradation elements as an opportunity to experiment a new 'aesthetic of leftovers'. In conclusion, the text argues to rethink the disciplinary status of the project and to adopt a new form of transdisciplinarity, defined as 'design of the unfinished', positioned between architecture, design, scenography, restoration, exhibition design.

In Chap. 2, Invernizzi deals with a state of the art on the design of the unfinished as a valid perspective to contemporary interiors, which allows to move away from the speed of a world that does not pay attention to details, history and memory of places. Indeed, a succession of unique microcosms shows an increasingly widespread design approach in Europe and around the World, that pursues different ways of dwelling the leftovers and living contemporaneity. Kindness in a new design, in a new teaching, in a new generation of designers committed to safeguarding the existing in its fragility turns out to consider the Design of the unfinished as the first step for a new discipline.

In Chap. 3, Di Prete investigates an increasingly widespread phenomenon of refunctionalisation that starts from interiors but also affects the surrounding urban environment and, sometimes, even the urban layout, based on the transformation of several buildings once used for different purposes, which today are rethought as cultural and creative hubs, places of exhibition and research. These actions have encouraged integrated public–private partnerships, often leveraging art and culture to activate the local spontaneous resources and, on longer horizons, to change the urban image (and imaginary).

In Chap. 4, according to Anzani, reusing urban leftovers can facilitate place attachment, intensified by awareness of the place history: as shown by research in environmental aesthetics, historical places allow people to feel a sense of continuity, arouse curiosity and increase motivation to discover the places forgotten past. Reuse interventions should return dismissed areas to new use possibilities, consistent with the place nature and symbolic significance, involving people in synaesthetic experiences and engaging atmospheres.

In Chap. 5, Lonardo illustrates the 'unfinished' as a didactic topic which in the last five years has been proposed in the Final Year Laboratory of the School of Design of the Politecnico di Milano. The students have been involved in design projects on dismissed territories and buildings of a different nature, linked by the condition of being Leftovers. First, the chapter deals with a reconnaissance on the topic; then, a set of strategies presented to the students is described as an example, as well as some of those provided by the students themselves.

In Chap. 6, Coralli examines the urban transformations under way in South Benin through the prism of heritage. While events, by definition ephemeral, have difficulty in reviving the colonial heritage and giving it meaning, initiatives led by artists' collectives periodically attempt to stimulate urban regeneration dynamics. The 'aesthetics of the leftover' suggested by Crespi (2018) poses an important challenge for the valorisation of innumerable urban spaces in sub-Saharan Africa with unfinished landscapes, derelict buildings, unbuilt spaces and abandoned building sites. The term wasteland (Amselle, 2005) is understood as an alternative place, an intermediate space where new artistic forms are produced and that, paradoxically, derives its vitality from ruins.

In Chap. 7, Matz presents Theaster Gates' built environment projects within the variegated geographies of the City of Chicago's South Side. Thaester Gates, who has built his early career within the arts, creates, owns, and operates a unique portfolio of projects. As an evolving collection of growing, interrelated work, they define a singular programme of what can be termed creative systems innovation. And within the context of this subject, the built-environment creative systems innovation of the unfinished.

In Chap. 8, Katoch tracks the history of the abandoned Dhan Mill Compound area in New Delhi, and the forces affecting its recreation as a creative hub. The narrative speculates on the isomorphism of such development projects, within the context of leftover spaces, across continents, driven by repatriated commonwealth communities, between India and the United Kingdom.

In Chap. 9, Crippa describes how the so-called 'urban remnants', with their suggestion of the unfinished, their versatile spaces open to welcome any story or activity together with their charm, fascinate visitors who are caught into a dramatic atmosphere leading to multiple interpretations. In particular, during the Milan Design Week over the past 15 years, urban remnants turned into the setting for some of the main events, recounting a series of stories that told their strong calling for theatre, through little set-up works becoming setting for unusual metropolitan shows/events.

In Chap. 10, Caramel rethinks the theoretical framework within which possible solutions to safeguard urban leftovers can be designed. As mentioned in the chapter, restoration is no longer a question of recognising the object, but rather of who recognises himself in that object. Therefore, the focus is not on objects, but on how they are perceived and experienced by people. On this basis, the contemporary ethics call for *negotiatory* conservation in which both voices of academics and of general public and future users must be heard.

In Chap. 11, Guglielmi defines leftovers as 'no-man's lands', which also evoke the metropolis 'non-places'. However, they contain a great use and integration potential that could be developed through the rules of self-similarity which are present in nature and art, as well as in the territory architectures. The Anti-nature crisis, that began around the 1920s and culminated at the end of the 1970s, can find its exit precisely in the theory of the Leftovers, considered no longer as chaos objects but, in their still unfinished dimension, as real models according to fractal rules.

In Chap. 12, Cristini reflects on the architectural 'leftovers' which constitute a focus for artistic research, facing the concepts of temporariness and precariousness, as constitutive elements of contemporary culture. The relationship with architectural 'leftovers' can go beyond a static vision of their recovery as containers, in the attempt to 'negotiate' with space in order to 'giving form to precariousness'. The chapter examines two case histories: the exhibitions of the Dialogos project (at Assab One, a former print shop in Milan, and at Vitrína Deniska a former shop window in Olomouc in the Czech Republic) and Sopra-Sotto (a former supermarket in Sesto Calende, in Italy).

In Chap. 13, Forlenza and Jennings discuss how the concept of leftover can be applied with interesting results in music. In fact, the idea of leftover turns out to underlie the musical discourse in three ways, which are all intertwined, while maintaining their own specificity. According to these three declinations, the leftover can be defined as a synonym of surplus, renewal, and imperfection, i.e. something which is left *unfinished*. To conclude, the Authors go back to Bach's *Preludes and Fugues*, readapted by pianist Brad Mehldau, as a clear example of all of three acceptations of the *leftover.*

In Chap. 14, Tognetti proposes a survey on the reasons why the idea of planning is facing a crisis and introduces the concept of 'Softlaw' as a new paradigm which could turn a contemporary project into a self-regulating device and stir things up in a field where everything is against planning. Besides, a Manifesto for responsible and creative Local Governance is presented, which lays the foundations for a disciplinary revision of the Italian Constitution based on art. 118 and aims at applying the subsidiarity principle. The essay deals with some aspects of today's Italian situation but tries to face the problems in an international dimension.

Luciano Crespi

References

Amselle JL (2005) L'art de la friche. Essai sur l'art africain contemporain. Paris: Flammarion
Bianchetti D (1985) Aree industriali dismesse: primi percorsi di ricerca. Urbanistica, 81
Bourriaud N (2020) Inclusioni. Estetica del capitalocene. Milano: Postmedia books
Chaix R (1989) Friches industrielles et réaffectations en Ile-de-France (Evolution 1985–1988). Hommes et Terres du Nord 4, 320–324
Crespi L (2018) Manifesto del design del non-finito. Milano: Postmedia books
Roberts P, Sykes H (2000) Urban Regeneration. A Handbook. London: SAGE
Secchi B, Boeri S, Brandolini S, Bianchetti C, Gabellini P (1984) Un problema urbano: l'occasione dei vuoti, Casabella, 505
Semi G (2015) Gentification. Bologna: Il Mulino

Contents

Part III Disciplinary Contaminations

Contributors

Anna Anzani Department of Design, Politecnico di Milano, Milano, Italy

Claudia Caramel Department of Design, Politecnico di Milano, Milan, Italy

Monica Coralli Laboratoire Architecture Anthropologie, CNRS, UMR 7218, Lavue, Paris, France

Luciano Crespi Dipartimento di Design, Politecnico di Milano, Milano, Italy

Davide Crippa Dipartimento di Design, Politecnico di Milano, Milano, Italy

Ermanno Cristini Artist, Varese, Italy

Barbara Di Prete Dipartimento di Design, Politecnico di Milano, Milano, Italy

Pier Francesco Forlenza Conservatorio di Musica G. Verdi, Milan, Italy

Eugenio Guglielmi Design Campus, Università degli Studi di Firenze, Calenzano, Italy

Didier Houénoudé National Institute for Art, Archaeology and Culture, University of Abomey-Calavi, Abomey-Calavi, Benin

Fiamma Colette Invernizzi Dipartimento di Design, Politecnico di Milano, Milano, Italy

Michael Giovanni Jennings Conservatorio di Musica G. Verdi, Milan, Italy

Mitaali Katoch Managing Director of Development; Company Associate, Charles Matz, New York, USA

Emilio Lonardo Politecnico di Milano, Dipartimento di Design, Milan, Italy

Charles A. Matz III The School of Architecture and Design, The New York Institute of Technology, New York, NY, USA

Roberto Tognetti Fondazione Riusiamo l'Italia, Borgomanero, Italy

Part I
Politecnico Approach

Chapter 1
Manifesto of Design of Unfinished

Luciano Crespi

Abstract The text identifies neo-nomadism as an issue characterising this beginning of the century and as a challenging field for the project culture. It calls for questioning the meaning of living in the contemporary world, in post-industrial cities, in relation to available spaces no longer used, which are named leftovers. Their regeneration project is described, which can assume their degradation elements as an opportunity to experiment new "aesthetic of leftovers". In conclusion, the text argues for rethinking the disciplinary status of the project and for adopting a new form of transdisciplinarity, defined as "design of the unfinished", positioned between architecture, design, scenography, restoration, and exhibition design.

Keywords Unfinished · Leftovers · Interior design · Exhibition design · Transdisciplinarity

1.1 Leftovers

In cities throughout the world, large and small, there are more and more spaces, both private and public, created in distant and recent years, which seem to have come to the end of their useful life.

In a text on the transformation of cities in the post-industrial era, the American economist Saskia Sassen defines as "improbable" the spaces whose future options are difficult to predict because of the speed at which today's "utility logic" changes. It is mainly for design or the art world that he believes he is able to imagine possible ways of use, however temporary: One "can't help but think that artists are part of the answer; whether ephemeral public performances and installations or more lasting types of public sculpture, whether site-specific/community-based art, or nomadic sculptures that circulate among localities" (Sassen 2006). Among the improbable

---

The chapter is a reworking of book "Manifesto del design del non-finito", 2018, postmedia books, Milano

L. Crespi (✉)
Dipartimento di Design, Politecnico di Milano, Milano, Italy
e-mail: luciano.crespi@polimi.it

L. Crespi (ed.), *Design of the Unfinished*, The Urban Book Series,
https://doi.org/10.1007/978-3-030-73457-2_1

spaces, we can place the numerous types of spaces that I have defined as *leftovers*. These are places which, having ceased to perform the function for which they were created, find themselves, without citizenship, as though in a state of suspension and waiting.

They are too unattractive, in terms of economic value, to be included in the investment programmes of large real estate operators, for whom the size factor often plays a decisive role. In the last few years, numerous interventions have been carried out all over the world to reconvert and reuse large abandoned industrial areas, through procedures that required considerable resources, with the involvement of a great deal of financial and institutional agents. In this scenario, small spaces and single abandoned buildings remain for years as shady areas within the urban system; disposable voids, to be destroyed; resorting, in cases where they are within urban areas of high real estate value, to their demolition and reconstruction, or renovation for speculative purposes. The issue of size is not always the crucial factor. The same fate is reserved for larger spaces. They survived *extinction* for reasons similar to those that determined the survival of some species in the selection process that took place in the living world. However, since they do not possess the necessary requisites to make their own redemption possible, because of their unfortunate location, the social conditions of the context in which they are placed, or anything else, they are condemned to a progressive degradation; and to become, in many cases, ruins.

The *leftovers* are not of such extraordinary historical and artistic value that they deserve to be restored and returned to their original condition, as is the case for monuments and certain buildings of worship or public interest which, while no longer fulfilling the function they once had, are preserved, even if only as "museums of themselves", eloquent witnesses to the history of which they were part. There is no time threshold able to determine precisely when an abandoned space may or may not fall into the *leftovers* family. It can only be said that they are more likely to be found within what has been achieved since the beginning of the twentieth century.

Lastly, they are too recent, still recognisable and not so shaken by the signs of time, to be able to assume the noble and romantic rank of the *ruin* (Brandi 1963), to which the territory can grant the chance to enter a process of progressive degradation and abandonment, which will lead it towards an inevitable end, after having represented, like ancient ruins, a form however melancholic of attraction. It is the Italian art historian Manlio Brusatin who reminds us that "wrecks quietly become a new construction" inhabited only by echoes, and that ruin is destined to fill "the future with its own past" (Brusatin 2000). In his engravings, Giovanni Battista Piranesi calls the tomb of the Metella family on the Via Appia, in ancient Rome, a leftover; as well as the tomb of the Plauzia family on the Via Tiburtina, and that of the second floor of the *Terme di Tito*. But they are ruins, condemned to live their condition of uninhabitable emptiness, *hinabitabilis*, in silence: matter that is only good for experts in the art of oblivion.

So, *leftovers*. Like the leftovers of lunch or dinner, which in consumer society have always been thrown away, and yet now it is beginning to be considered immoral, as well as senseless, not to reuse them. Like the leftovers of time, which make those who can dispose of them happy. Or leftover fabric, with which you can, with a little creativity, invent new and unimaginable outfits. Or the leftovers of marble processing, with which design objects can be created. Or the *scraps* from woodworking, used by Italian designer Marco Zanuso junior to create tables. Or the bottles and containers of detergents used by Massimiliano Adami, Italian designer and artist, to create his *modern Fossils.*

Leftovers such as *briccole,* ("Briccole" are the wooden posts used to mark the berths and navigable waterways through the lagoon of the city of Venice) were used by Alessandro Mendini: "When it performs its function in the waters of Venice, the *briccola* is a poetic and light object, which is reflected in the water in a kind of slight evanescence. When it has completed its primary task, when it is removed from its place to be replaced, then the *briccola* remains a long pole on the ground, where wear and tear and time have worn down the wood, creating a hard, uneven crust. Nevertheless, that wood has a rich history, and it is fascinating to want to bring it back to life. In my case, I wanted to achieve a gentle image by reversing the harshness of the trunk. The *Unicorn,* expressed in various forms, is an ironic-looking stool, like a good luck toy. A wooden toy, in fact, carved in a primordial way, as the carpenter Geppetto did with Pinocchio. A toy for adults, with blue, round Murano glass eyes and a golden bronze horn".

Aristotle wrote, "a proverb is a remnant from old philosophy, preserved amid countless, destructions, by reason of its brevity and fitness for use". Unlike waste, which is born with the stigmata of the defective being, *leftovers* are not to blame, unless the blame is for being useless, until someone discovers and reveals their hidden value, which is also represented by its ability to elicit, in some cases, "automatic empathic reactions", similar to those that *neuro-aesthetics* considers being provoked by a landscape, although in this instance the weight of contextual, experiential, cultural and cognitive factors appears far greater. The photo of an old house, without any particular beauty, leads Roland Barthes to say that "it is there that I would like to live", because, writes the Italian psychoanalyst Vittorio Lingiardi, "of a place, what attracts Barthes is habitability", something that has to do with the idea of *consonance*, "a sense of emotional harmony with the environment", dependent on time and memory. *Leftovers* can become part of the objects and landscapes whose structure is able to exert some form of attraction on people. Because, for Lingiardi, "the identity of places is also the result of the gaze they have received and, often, of human work" (Lingiardi 2017).

In the book *Riusiamo l'Italia Da spazi vuoti a start-up culturali e sociali [Let's reuse Italy—From empty spaces to cultural and social start-ups]* it was estimated that if, in Italy, we consider former depots, slaughterhouses, schools, churches, barracks, as well as shops and houses, there are about six million abandoned or unused spaces (Campagnoli 2014). There is a lack of precise data on the scale of the phenomenon of the abandonment of spaces throughout the world. However, there is an increasing number of photographic evidence on the Internet regarding places that can be classified as *leftovers.* There are also sites such as "30 abandoned places in the world. From India to New York: Ghost Cities and Sunken Villages", or "47 Abandoned Places that will give you Goosebumps", or "45 Abandoned Places Around the World That Are Eerily Beautiful." The idea is also emerging that when some of these places have particular characters, due to their symbolic or iconic value, they can be transformed into a melancholic circus attraction, like the village of Goussainville in France, a sort of transposition into reality, for those who love splatter, and the murky atmospheres of *Sin City;* or like the now famous Hashima, an abandoned island opposite Japan, which was also used as the film set of the 007 film *Skyfall.*

Their variety is such as to present an analogy with the world of living forms. Biologists have classified living organisms according to an organisation on different levels, which includes, in the case of the animal kingdom, the type, class, order, family, genus and species. In the case of *leftovers*, we could consider the type (*phylum*) in relation to several spatial characteristics that unite them, in the same way as, to return to the classification of the animal kingdom, all animals with an internal support structure called "notochord" belong to the *phylum Chordata.* It is not a question of taking up the notion of the *type* elaborated by the tradition of theoretical studies that includes, among the most authoritative, Argan, Caniggia, Aldo Rossi and Martí Arís (Caniggia and Maffeis 1982; Martí Arís 1993). If not for the part that attributes to the type qualities indifferent to the function that the space performs. In the case of leftovers, the "classic" typological approach would take for granted the total heterogeneity and randomness of their spatial configurations, often born not as a resumption or evolution of a recognisable type, as it is known, not as a variation of identity, but as an entirely functional and pragmatic response to a need for use.

Spanish architect Carlos Martí Arís can be taken to mean that "the form acquires a more stable character than the specific activity that generated it" (Martí Aris, 1993). Recognising the "character" of the leftover is the necessary premise for carrying out its regeneration, without erasing the signs of its deterioration and those representing its spatial logic. Both belong to the ontological nature of the leftovers.

We can talk about class in relation to the function performed previously, before the divestiture. It defines the degree of belonging to a role exercised over time, which represents the content of memory and which cannot be overlooked when assessing the different perspectives of reuse of space.

The regeneration of a *leftover* necessarily implies a change of nature, something that resembles the changes of state in chemistry, in order to update it and make it attractive to the new demands of use of space as expressed by the contemporary world. *Leftovers* have their own history, whether long or short, rooted or not in the collective memory of the place to which they belong. This history is an "integral part" of the *leftovers;* it constitutes its identity, its reason for belonging to the place and for survival, even after the function for which it was made has been fulfilled. There are no automatisms able to predetermine and condition the passage from the old function to the new ones. There are reasons for "good manners". This must apply in all instances, but especially in those, where space has played a role of great symbolic value in the previous life compared to the territory in which it is located. Consider, for example, former churches or places of worship in general, for which the search for a new function cannot be separated from the previous one (Capanni 2019).

One can speak of *order* in relation to its location. The presence and conservation of leftovers is a choice that is aimed at maintaining a sort of biodiversity with regard to the existence of post-industrial city environments. Just as the *Third Landscape* constitutes a territory—writes Gilles Clément—"for the many species that cannot find space elsewhere" (Clément 2005), the set of *leftovers* represents a territory embedded in the porous structure of the contemporary city for types of spaces that would not find citizenship elsewhere. They are, in a certain sense, "residues" that arise from the abandonment of an activity. They can become "reserves", being, again for Gilles Clément, "sets judged as fragile or rare, rich in an endangered diversity".

Of *families*, in relation to their physical characteristics, the use of materials and the type of degradation, the *leftovers* present unique and unimaginable characteristics, due to the state of abandonment in which they are found, which has produced alterations in the materials it is made of, capable of generating real new chemical formations, surprising chromatic effects, rippling of surfaces, folds, distortions of meaning, voids and spontaneous forms of vegetation in the interstices. This is a crucial feature of the *leftover*, which the regeneration project cannot disregard.

Of *genus* and *species* in terms of size and character, both of the internal space and of its container.

A certain component of the arbitrary nature of the attempt to apply a classification system valid for living beings to the inanimate world of environments, which we have called *leftovers*, is justified by the need to assign a systematic character to the work of research and design experimentation, in order to give the project of *leftovers* the more general value of a *style of thought* concerned with measuring itself again some crucial aspects of contemporary life. Although it has its own specificity, in relation to the geographical context in which it is situated, each leftover must be taken as a paradigm of a different way of interpreting the phenomenon of the disposal of spaces in the light of their possible reuse.

1.2 Urban Leftovers

It cannot yet be said that the true phenomenon of the abandonment of urban open spaces exists. The system made up of streets, squares, voids and widenings, is so vital for the operation of the city, a delicate organism based on permanent flows of people and things, that it cannot be deactivated except for very short periods of time or due to accidental causes. It would therefore seem inappropriate to speak of leftovers in relation to urban open spaces. There is the phenomenon of degradation to which portions of this system are often subjected, due to the underestimation by public bodies of the importance of what was once called urban decorum. After the presumed death of public space due, according to the American sociologist Richard Sennet, to the irruption of intimism in everyday life, whose effect would have been to push people to seek in the private sphere what is denied of them in the public sphere (Sennet 2006), we have witnessed its rebirth in recent years. Having lost the character of a specialised place, contemporary urban space is now required to accommodate the multiple methods of self-consumption on the part of the user, allowing everyone to build a sort of personal palimpsest, on the basis of which they can also interact with the devices present on the site, modifying it. And also to enter into a more "intimate", empathetic relationship with its users, so as to make them somehow protagonists, albeit in very different forms than in the past. The multiplication of these questions has not been matched in the last century, except sporadically and in certain countries, such as Denmark, Spain and France, by a widespread redevelopment of such spaces. In the early years of this, in South America, in some East Asian countries, such as China and Korea, and in the United States, the redevelopment of "urban interiors" has multiplied.

Urban leftovers can therefore be called places, streets, squares and widenings, which do not have the character of spaces that are hospitable, welcoming and endowed with civil eloquence, but which some of them possessed at the time of their creation and which the lack of an effective maintenance and innovation policy has erased over time. They constitute a *leftover* not because they no longer perform the task assigned to them, but because they perform it inadequately, compared to other open spaces that are able to take care of those who use them and comply with the demands of use made by new, contemporary users.

The *urban leftover* lives in a singular state of suspension, waiting to find not a different function more appropriate to the present time, since that function is irreplaceable, but a different way to perform it. In the (perhaps not too distant) future, it is possible that radical changes in the ways people are moved, for the time being in an experimental state, such as the flying taxi, which Uber believes may come into operation in 2023, or proposed by science fiction films, may have repercussions on the layout of the urban open space, producing actions for the disposal of some of its parts, as they are no longer necessary for the sorting of the flows they carry out. One may be faced with a phenomenon whose size and nature are difficult to assess at present. Nevertheless, that could have parallels with the decommissioning of some urban railway tracks and their conversion into pedestrian spaces, as was the case with the New York *High Line*.

1.3 Neo-Nomads

Leftovers can be an extraordinary resource, not only because they are available to carry out new functions and to host, even temporarily, cultural, residential and social activities, able to respond to non-programmable demands for use, determined by occasional states of need, without incurring further land consumption, but also because they are often located in central, or at least not peripheral areas of the city, where they play a role of high symbolic value. They are custodians of memories and human stories otherwise inexorably destined to be lost, or entrusted only to the feeling of melancholy felt in the face of past testimonies and events, no longer able to exert any influence in the present time. Reusing *leftovers* is not only ethically sound. It represents an unprecedented opportunity and a unique challenge for the project. The *leftover* lends itself to redemption operations aimed at offering multiple forms of hospitality to contemporary travellers, "border crossers", neo-nomads, migrants, in a sort of passing of the baton, from the previous stories of which those places have been the backdrop, to the new stories that have yet to begin.

Even before the refugee phenomenon took on its present proportions, there were many signs that the first forms of mass migration would spread across the planet. The 2017 UN International Migration Report estimated that some 258 million people have left their countries of birth and now live in other nations—an increase of 49 per cent compared to 2000. Even though they are different in nature, the two phenomena are indicative of the same indisputable trend: the appearance on the contemporary scene of a new social actor, defined by some as a wayfarer of modern times, by others as a *neo-nomad*.

Michel Maffesoli, professor emeritus of Sociology at the Sorbonne in Paris, was among the first to deal with the subject in his book, *Du nomadisme: Vagabondage initiatique* (Maffesoli 1997) and to argue that the nomad can be considered one of the emblematic figures of our time. In this case, nomadism is seen as a form of wandering, something embedded in human nature and which can represent a kind of catharsis, an expansion of the self towards something greater, capable of encompassing the earth, the world and others. It is an interpretation that undermines its explosive and political meaning in favour of one more connected to the vague idea of social neo-solidarism, which finds in the web the place that makes new ways of interconnection possible.

Because of the profound changes that have taken place in the contemporary way of living and dwelling, concerning almost the entire planet, for the Italian philosopher Umberto Galimberti, all that would be left to man would be "the destiny of the 'wayfarer', who, unlike the 'traveller' who journeys along the road to reach a destination, embraces from time to time the landscapes he encounters on his way, and which for him are not places of transit waiting for that place, Ithaca, which makes every land a simple stop on his way back" (Galimberti 2000).

For Arianna Dagnino, Italian writer and scholar of changes in collective behaviour, the new nomads are the great explorers of geographical and mental, physical and virtual frontiers, possible city dwellers who could be born and die in the course of a single day. About 20 years after her study on the *new nomads* (Dagnino 1996), Arianna Dagnino upholds the need to adopt a new approach to the problem of the explosion of migratory phenomena, able to overcome the major limits of the policies used towards them, based on the presence of two specular cultural attitudes, with one defined as an "assimilationism" and the other as "multiculturalism". She defines it as a transcultural approach, as "a critical perspective that understands cultures as dynamic processes affected by fusions and confluences" which "can therefore also represent an alternative model of identity-building, which develops at the intersection with other cultures and pushes towards a dimension that goes beyond any specific culture (…) Transculturality, instead of focusing on polarities and differences, favours intersections, elements of commonality and shared initiatives". This is why she attributes to some spaces of the past, for example, caravanserais, considered "fertile crucibles of interculturality", the property of producing relationships which, although temporary, are able to constitute forms of "stable anchorage" for those involved in the different modes of contemporary nomadism. She believes that the task of reconstituting occasions for re-knitting the canvas of social relations can be entrusted to these, and more generally to the places that she defines as "third spaces", where living, commercial, cultural or even "contemplation" functions coexist (Dagnino 2016).

Although starting from a new disciplinary perspective, that of art criticism, the Frenchman Nicolas Bourriaud reaches the same conclusion about the failure of post-modern multiculturalism, the cause of which is due to its functioning on the basis of belonging, inevitably destined to create "ethnic roots". If "the immigrant, the exiled, the tourist, the urban wanderer" are the dominant figures of contemporary culture, and if one intends to operate in a situation defined by migratory flows, by planetary nomadism and by the globalisation of financial exchanges without having to endure it, it is necessary to "develop a nomadic thought that is coordinated in terms of circuits and experimentations, and not of permanent installation. We oppose the precarization of experience with a resolutely precarious thought that is placed and inoculated in the same networks that suffocate us". A thought that replaces the idea of identity rootedness with the notion of *rooting*, which implies the abandonment of "exclusive disciplines" in favour of "a nomadic party, whose main characteristic lies in inhabiting existing structures: accepting the fact of becoming the tenant of present forms free to modify them more or less deeply"[1] (Bourriaud 2014). The idea of inhabiting existing structures is here referring to the context of the art system, with its insistent remixing of the notion of space-time on which the work is based. However, the more general sense of these considerations is much more than a seductive intellectual appeal, also for the project.

The Catalan designer Marti Guixè, with his proposal for a herbal *relaxation* pill, makes a brilliant and subtle provocation about the need to take into consideration the issue of mental and physical adaptation of workers to their condition as contemporary nomads. As in all the Catalan designer's works, the conceptual component prevails over any other: we are at ground zero in terms of writing, at the extinction of form in the face of the power of content and its metaphorical meaning. This goes to the root of the reasons for living in contemporaneity, and gestures, such as travelling without luggage, which are able to extend the concept of home to the rest of the world, where nobody, Guixé reminds us, moves from one room to another holding a suitcase or a bag (Crespi and Di Prete 2021).

For Stefano Boeri, Italian architect and scholar, "never before has the European city become a sort of large camp; a home for temporary stop-offs and mobile living projects" (Boeri 2011).

If the great challenge of these years, and probably of this century, first of in terms not only of politics but also for the project, is to give credible and practicable answers to the demand to live generated by the non-momentary phenomenon of the displacement of millions of human beings from one geographical area to another—even if for very different reasons—it seems evident that the solutions proposed by design culture, and not only, developed during the twentieth century, are no longer available.

[1] Free translation from Italian.

1.4 Inhabiting the World

Even the most sophisticated and profound reflections formulated on the notion of space to be inhabited appear inadequate: from Walter Benjamin's reflections on the *interieur* as "not just the universe but also the étui of the private individual", a place where one is "cradled in one's own illusions" and from which the phantasmagoria about it originates: "For the private citizen, it represents the universe, in which he collects the distant and the past, and his living room is a stage in the universal theatre" (Benjamin 2002); to those of Gaston Bachelard, on the house as a "corpus of images that provide man with reasons or illusions of stability" (Bachelard 1957); in addition, to all those belonging to the tradition of architecture developed during the twentieth century. The same notion of *existenz minimum*, experimented within the first half of the last century and aimed at guaranteeing a living space, however minimal, even to the most disadvantaged strata of the population, seems today impractical in the same terms. If at the time the idea was to guarantee the well-being of inhabited environments through a codified quantity of spatial and dimensional requirements, today the project is required to give answers to needs that extend far beyond these aspects. More than 30 years ago, Alessandro Mendini, one of the masters of Italian design, had foreseen everything. For him, the "serial standards" typical of contemporary society are merely a simplistic attempt to respond to mankind's elementary survival needs. For him "man urgently needs other kinds of survival, a subtle survival: instead of a living room, kitchen, toilet and bedroom, we all need new types of rooms and cabins: perhaps rooms for swimming, for keeping flowers in, for telecommunications, for reading books" (Mendini 1982).

Published for the first time in 1974, the book *Il settimo uomo (The Seventh Man)*, written by John Berger, a great critic and author who died in 2017, was published in Italy in 2017, with photos by Jean Mohr: an epic tale of emigration and suffering. It is from that experience that the condition of contemporary living is described by them in this way: "The house is a microcosm that functions as a model to interpret one's daily experience and to structure and project it externally. However, in a world that is moving, made up of global workers, emigration and exodus, it is appropriate to reflect on what new meanings the home can acquire. The theory is that rather than physical space, it increasingly appears in the form of routine clusters and practices that are exportable elsewhere, i.e. opinions, clothing styles, etc. Home as habitus, incorporated practices, in the words of the French anthropologist Pierre Bourdieu, as performance. Narrations about the home become the ideal research context to discover how individuals build *moving homes* of various kinds, 'cognitive homes' in transit"[2] (Berger and Mohr 2017).

Starting from a reading of the twenty-first century as the century that highlighted "the crisis of the city and consequently the crisis of the project as an instrument of order, progress and beauty", the Italian design critic Andrea Branzi sees as the only perspective for the project the need to go back to the deepest anthropological roots of man, to return "to that primordial condition from which the human environment

[2] Free translation from Italian.

began its long journey" (Branzi 2016). It is a seductive but also disturbing perspective, discreetly apocalyptic and disciplinarily irrelevant, especially if his prophets were, as he suggests, *hikikomori*, *plumbkingers*, *owlingers*, *voguers* and *parkourers*.

It is necessary to think of the city's interiors as a laboratory in which to experiment new ways of living spaces, freed from their boundaries, repeatedly staged and in which private and public elements mix and deconstruct.

1.5 Transdisciplinarity

The culture of design, and in particular of interior design, seems equipped, for history and attitude, to play a leading role in this perspective. It adopts an original design conduct, different from not only that of the disciplines of restoration or building renovation, but also of furniture as it is commonly understood. The challenge is to experiment with interventions capable of assigning new possibilities of use to the *leftovers*, using temporary and reversible equipment, provided that it is consistent with the nature and *soul* of the place, in order to encourage their reintegration into the social fabric and the enhancement of their symbolic content. Through a transdisciplinary approach, it is a matter of welcoming into the project as a "gift" the elements of degradation, "wounds", wrinkles, creases, folds present in the existing work, to translate them into a syntax aimed at giving shape and meaning to the internal and external environment with the inclusion of "additional components". The project of *leftovers* acquires, in this sense, the more general value of experimentation of a borderline discipline, placed between design, interior design, arts, restoration, exhibition design, set design, cinema and photography, with the aim of inventing a "new tradition", based on what can be defined as the "aesthetics of leftovers". This could be defined as the *design of the unfinished*, understood as a way to attribute to these abandoned environments, whatever their function, a representative *character* of the conditions of temporariness, precariousness, trans-culturality, typical of the century that has just begun, through the use of its own aesthetic code, and recognisable, as has happened in other moments of history.

In the world of art, the fifteenth century is represented by the works of Leonardo, Brunelleschi and Leon Battista Alberti and by the rediscovery of the ancient, the values of proportion and measure, mathematical and geometric precision. The sixteenth century is characterised by continuous research and experimentation, guided by a strong sense of uneasiness, by the lack of points of reference also with regard to past history and by the "centrality of doubt". Michelangelo was a masterful interpreter of these values, especially with the choice, in the final part of his existence, of the unfinished. For the great Italian art historian Giulio Carlo Argan, "modern critics have seen in Michelangelo's unfinished sculpture the supreme moment of art that transcends its technical limit (…). Since then, with the mannerists and then with the romantics, all art was constitutionally unfinished; and, going beyond its disciplinary limit, it challenged the prejudice that the finite was the necessary connotation of value" (Argan 2005). We do not yet know what historical events will mark the twenty-first century, but what happened in its first 15 years was to tell Luciano Canfora, an Italian historian and a profound connoisseur of classical culture, that, thanks to the unstoppable phenomenon of migratory movements and the spread of new forms of slavery, the problem facing the century is "rebalancing the unjust division of wealth as soon as possible. Without this, the conflict for survival (in its most varied forms, including the unliveability of the metropolis) will be the dominant feature of the decades that lie ahead" (Canfora 2017).

The idea of regenerating *leftovers*, according to the approach that has been suggested, appears—also because of their location in non-peripheral and socially non-degraded areas—to be a credible perspective from the point of view of the cost-benefit ratio: it costs little and requires short implementation times. It allows a true integration of the people welcomed, avoiding the creation of new ghettos. It is reversible so that it can accommodate changes in "utility logic".

1.6 Works

There are recent experiences of reuse of unused spaces that are indicative of a new style of thinking that is spreading in different parts of the world. Different from that experienced in some countries, especially in Europe, based on the assignment of the abandoned space to groups of artists, cultural associations, groups of citizens, to whom the task of temporarily managing it is entrusted, while waiting for it to be assigned a new function. It is an operation often carried out without a proper planning strategy, and entrusted mainly to *bottom-up*, social sharing processes, as in the case of the *NDSM werf* in Amsterdam. Or Hamburg's *Gängeviertel,* just to give a few examples.

Some of the experiences contain *in nuce* individuals of a possible paradigm shift in the direction of non-finite design. Studio R-ZERO's 2014 project to transform an abandoned twentieth century office space in Mexico City into an office space takes on these objectives: "It's about creating architecture without building. It seems an almost impossible goal, but one that is effective and innovative. The clarity of the concept is what determines the potential of a visually attractive project, transcending any physical element, which makes the project a reality". The building was designed to house several homes, but over its long life it has performed different functions, from the headquarters of the United Mexican Booksellers to a pastry workshop. When it was built, at the beginning of the twentieth century, the area was mainly dominated by the presence of country houses. The subsequent growth of the city and the building's proximity to the centre caused it to be abandoned. The previous experiences are precisely what gives it a distinctive personality and makes its interior space unique. As with a real individual, the designers write, the building acquires its character through its experiences, almost empirically, and therefore its distinctive signs, scars, and wrinkles are not erased from the project, but form part of the way space expresses itself when assigned to new use, and they nurture the plot of another chapter of its rich life.

Antivilla is a project by Berlin-based studio Brandlhuber + Emde, Burlon, built in 2014, for the redevelopment of the former German Democratic Republic's Ernst Lück lingerie factory in Krampnitzsee, southwest of Berlin. For the designer, "here the raw material does not represent an idea to which it is subordinate: rather it highlights its ability to create situations and scenarios of use. It is not a question of 'straightforwardness' of the material, but of the material as a pure state of aggregation". The façade was clad with sprayed concrete. The non-load-bearing walls were removed, to create a unique space, and a 20 m^2 functional core was installed on both floors to contain the staircase, the bathroom, a kitchenette, a fireplace and a sauna. It is possible to divide the space into zones, according to climatic requirements, using PVC curtains. In summer, the curtains can enclose a bedroom measuring 10 m^2 within a large open space of 230 m^2, while in winter the heated area is reduced to 60 m^2 and the bed is moved near to the fireplace. The existing roof was replaced by a flat concrete slab. The windows, up to five metres in size, were created with the collective work of friends who hammered the wall to obtain openings that allowed a view of the nearby lake.

St. Miquel 19, by Carles Oliver, built in 2016 in Palma de Mallorca, cost 18,000 euros. Of this, 12,000 euros was for the plant engineering part and the rest for the interior redevelopment. The initiative's focus on the theme of sustainability generated a real expressive language, based on three principles: 1. *Not to do*. Not doing, as the best way to do. 2. Domestic archaeology. To demonstrate how the use of reduced means has been the basis for the construction of our cities for centuries. 3. *Back to Arch.* A way to open spaces without adding any new material. Opening a door is not the same as building a door. The project preserves both the materials (such as stone and wooden beams) of the existing work, and the construction elements, such as doors and windows, and recomposes them as fragments of a new aesthetic code by Paulo Moreira, a young Portuguese architect. These are a café in A Sandeira, Porto, in 2013 and a brewery in Musa, Lisbon, in 2017. They are distinguished by the intelligent balance between the new furnishings introduced, and the elements found: the former using materials recovered from other buildings in the city and extending the granite slab flooring inside to create continuity in the space. In addition, the second—housed in a former warehouse with an uncertain destiny, as it was threatened by a possible real estate intervention that would see it being demolished—through a calibrated use of the interventions, a flawless design of the furnishings, the judicious use of resources also for the completely exposed plant engineering.

Between 2008 and 2013, the Italian photographer Andrea Di Martino photographed 70 deconsecrated and re-purposed churches, in many cases following an approach based on the use of minimal additional components.

This is only the beginning; these episodes occurred almost simultaneously in different geographical contexts, even before there was a conscious adherence to a thought that legitimised them, to a new paradigm. They were created rather as an inalienable need to measure oneself against some of the most problematic constituent traits of the contemporary world. These are interventions in which an approach that belongs to the culture of architectural design still prevails, based on the idea of "long-lasting" and irreversible transformation.

The design of the unfinished adopts a design thinking based on the judicious use of resources, the recourse to reversible interventions, the enhancement of the elements of degradation found in the places on which one has to operate and cannot disregard the formulation of new aesthetic codes, that is, "aesthetics of leftovers".

1.7 Teaching Experiments

During the teaching activities of the Interior Design Laboratory carried out in recent years, within the Design School of the Politecnico di Milano, this approach has been trialled, applying it to different types of *leftovers*, located in the territory of Milan and northern Lombardy.

Avanziuno: former drinking water pavilion at Trotter Park, Milano (in collaboration with the Municipality of Milano).

This is a "classroom" environment, whose beauty is bestowed not only by its grandeur but also by the existence, on the lower level, of what remains of the tanks once used to contain the water on which checks were carried out. It is almost an archaeological excavation. The need to preserve both the spatial quality and the traces of the former function, imprinted on the walls and ground, was met in the project by Federica Cocco and Luca Cotini with the introduction of a system of industrial scaffolding designed to provide suspended spaces for theatre, exhibition and video art activities. Detached from the walls, the system goes along with the choice of introducing the new one as an addition to the existing one, keeping its distance, not affecting the substance and the memory content that the *leftover* one preserves. In cases such as this, the system is reduced to the minimum necessary, with the use of space only for certain periods of the year—those that are less cold.

Photo: Federica Cocco

Figures 1.1 and 1.2

Avanzidue: former depot of Ferrovie dello Stato, Milano (in collaboration with the Municipality of Milano).

The space is located in an area undergoing a major transformation, at one of the stops of the tram line and near the former Lazaretto di Milano, built between the end of the fifteenth century and the beginning of the sixteenth century outside Porta Orientale, as a shelter for the sick during the epidemics, now characterised by the presence of a multi-ethnic population. Wedged into the drop between Viale Vittorio Veneto and the ramparts of Porta Venezia, the space has a completely underground blind front, is illuminated by openings on the other side and zenithal skylights and has a high level of humidity. This suggested the choice to introduce functions that require people to stay for just a short time, as in the case of Giuseppe Addesso and Helga Aversa's project, which allocates the space to waiting for the tram. The discovery of various types of objects inside has led to the choice of also furnishing the environment with these recovered items. The use of a language based on paradox and nonsense represents a rhetorical device to highlight the change in the nature of the

Fig. 1.1 Former drinking water pavilion, Milano

Fig. 1.2 Former drinking water pavilion, Milano

place and its regeneration, from a warehouse to a crucial meeting place for lives *en passant*. This case has a paradigmatic value, as it poses a series of problems common to many of the *leftovers*, concerning the hygienic-sanitary aspect and environmental comfort. These can be solved with a view to the temporary reuse of space.

Photo: Giuseppe Addesso, Helga Aversa

Figures 1.3, 1.4, 1.5, 1.6, 1.7, 1.8 and 1.9

Avanzitre: former school in Via Pianell, Milano (in collaboration with the Municipality of Milano).

This is a former primary school of the early twentieth century, on which some renovation work had been started at the end of the last century, which was never completed. The building appears to have been badly damaged by time, and therefore requires preliminary intervention to make it safe and restore some of the floors. The Municipality of Milan asked for temporary accommodation, to be allocated to people with serious housing problems, and collective spaces to serve the neighbourhood. The project by Ludovica Squillacioti and Giulia Tofi also involves the introduction of laboratory spaces and small commercial spaces. The different layers of materials,

Fig. 1.3 Former depot of Ferrovie dello Stato, Milano

Fig. 1.4 Former depot of Ferrovie dello Stato, Milano

Fig. 1.5 Former depot of Ferrovie dello Stato, Milano

Fig. 1.6 Former depot of Ferrovie dello Stato, Milano

Fig. 1.7 Former depot of Ferrovie dello Stato, Milano

Fig. 1.8 Former depot of Ferrovie dello Stato, Milano

Fig. 1.9 Former depot of Ferrovie dello Stato, Milano

present in the interiors, have been conserved and assumed as peculiarities, except in the accommodation, where measures have been taken to make these areas comfortable, albeit based on an idea of comfort different from that linked to certain clichés about living. Although this is a rather anomalous case of the regeneration of *leftovers*, due to the need to integrate the design approach with structural interventions in order to make it safe and usable, the interest lies in the fact of having challenged the idea of design of the unfinished, also in the regeneration of environments intended for residential activities, albeit of a particular type.

Photo: Mattia Bianchi.

Figures 1.10, 1.11, 1.12, 1.13 and 1.14.

Avanziquattro: Former depot of Ferrovie dello Stato, Varese (in collaboration with the municipality of Varese).

Located near Varese railway station, the former depot is a large, unique environment, with wooden trusses; a charming space, evocative of some ancient spaces of industrial origin. The materials worn out by the years, abandoned within, tell the

Fig. 1.10 Former school, Milano

Fig. 1.11 Former school, Milano

Fig. 1.12 Former school, Milano

Fig. 1.13 Former school, Milano

Fig. 1.14 Former school, Milano

story. Not only the photos that portray them but also those of scraps, cracks and signs of time are already designed. Reading them results in the idea, contained in the project by Gennaro Merolla and Federico Nunziata, to introduce a *Warehouse*, for contemporary art, as a landing point for the dense network of exchanges that animate the station. To that we can add a particular type of environment, the *speakeasy*, which refers to an experience in use during Prohibition in the United States, and which takes on an allegorical value, a reminder of a place that is a symbol of transgression and rebellion. This is undoubtedly one of the most eloquent cases of *leftovers*, with enormous potential for reuse, due to its location in the territory and the scenic qualities of the interior environment.

Photo: Michela Albergati, Federica Borrello, Elena Grisa, Gennaro Merolla, Federico Nunziata, Giulia Turati.

Figures 1.15, 1.16, 1.17, 1.18, 1.19, 1.20, 1.21, 1.22 and 1.23

Fig. 1.15 Former depot Ferrovie dello Stato, Varese

Fig. 1.16 Former depot Ferrovie dello Stato, Varese

Fig. 1.17 Former depot Ferrovie dello Stato, Varese

Fig. 1.18 Former depot Ferrovie dello Stato, Varese

Fig. 1.19 Former depot Ferrovie dello Stato, Varese

Fig. 1.20 Former depot Ferrovie dello Stato, Varese

Fig. 1.21 Former depot Ferrovie dello Stato, Varese

Fig. 1.22 Former depot Ferrovie dello Stato, Varese

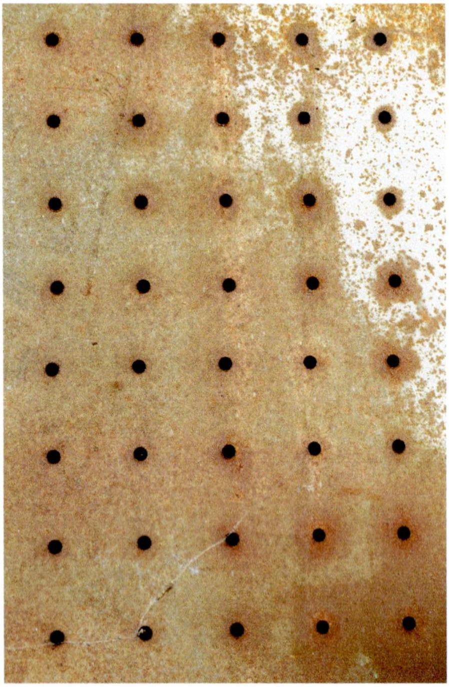

Fig. 1.23 Former depot Ferrovie dello Stato, Varese

Avanzicinque: Former church and commercial kiosk in Baranzate, province of Milano.

Located a short distance from each other, the small deconsecrated church and the octagonal kiosk, created to house drinks and food sales activities, despite not having recognisable architectural qualities, can lend themselves to a very ambitious redemption operation aimed at introducing a device capable of playing the role of new urban centrality. Letizia Agosta, Aurora Antonini and Francesca Spina's project takes its cue from the multi-ethnic character of the Municipality of Baranzate, to propose a "quadrilateral of cultures", based on a path able to correlate the two *leftovers* with other crucial places, such as the Library. The special geometry of the kiosk becomes a pretext for assigning to it an iconic value, as the result of a game of references to some figures, such as the baptisteries, of great significance in the history of architecture

Photo: Letizia Agosta, Aurora Antonini, Francesca Spina

Figures 1.24 and 1.25

Fig. 1.24 Former church and commercial kiosk, Baranzate

Fig. 1.25 Former church and commercial kiosk, Baranzate

1.8 Design Experiments

Competition project for the regeneration of the former Military Cold Store of Cuneo, 2020

Project: Luciano Crespi. Collaborators: Francesco Antonelli, Lorenzo Bressan, Marta Crippa, Lucia Ratti, Mirco Sturlese and Alexa Tamburrini. Pictures of Angelo Ariti

The former military cold stores are a case in point. Time has left its mark on this architectural work. Nature has taken back what man had taken from her and now *inhabits* the space, which has been reduced to just an enclosure. What appears to us is an image of rare symbolic power. A sort of allegory of the narratives in which the city of Cuneo can be recognised, like the one about the "seven sieges". Even today it's easy to read the plot, already present in the views of the *Teatrum Sabaudiae*, of an urban device enclosed between the walls, now replaced by the railway track, and the Gesso river: it is as if everything were born at that point, at the confluence of the two rivers, a sort of *cosmological principle*, where everything has its origin. Here stand the former cold stores, as though guarding not only a precious piece of the city, but the whole city itself, as it unfolds towards the south-east, and seems to gradually lose intensity, rarefy and reconcile with the landscape until it becomes confused.

The project assumes this condition as an unavoidable fact. It maintains the existing wall on Via Sette Assedi and Viale Kennedy in its current configuration, not as a "mask" or skin, but as the *substance* of the new life for which the place is designed: a wall characterised by a precise rhythm marked by pilasters, placed three metres apart, by "lunette" openings, by "wrinkles" created over time. It preserves "the room of twenty columns" as an extraordinary triple-height space to be used for exhibition events, whose last room, reserved for the rarest works, must be conquered, as in a new, exhilarating siege.

On the ground floor and the first floor of the long wing are the rooms for other uses, in spaces that preserve the character of the existing place, with the office and service areas added as "coloured boxes", equipped with the necessary comfort and microclimate, within a sort of artificial landscape. On the first floor, where the training room is also located, it looks like an eloquent volume, characterised by the pitched configuration of the ceiling, with metal trusses and natural light filtering through the windows, positioned to connect the existing walls with the new roof.

Instead of the current car park in Via Sette Assedi, a *Xenia* is planned, a hospitable place where people can stop on the long "street bench" at the foot of the wall, as in the tradition of many historical buildings, and meet the *other* in the shadow of the seven hornbeams, a symbol of vitality and freshness, placed between water signs that complete the design of the square, renewing in an allegorical way the link that the city preserves with its rivers.

Figures 1.26, 1.27, 1.28, 1.29, 1.30, 1.31, 1.32 and 1.33.

Fig. 1.26 Compatition Project, Rigeneration of ex military refrigerator, Cuneo, 2020, View of the building

Fig. 1.27 Compatition Project, Rigeneration of ex military refrigerator, The inside of the building

1.9 Conclusions

The twenty-first will be, in the field of the project, the century destined to measure itself against the theme of temporariness and precariousness, also from the point of view of aesthetic practices.The design of the unfinished is inscribed in this perspective and requires the development of its own syntax.

As far as the contents are concerned, it involves the use of figurative language: metaphor, allegory, hyperbole, in order to give a symbolic value to the solutions

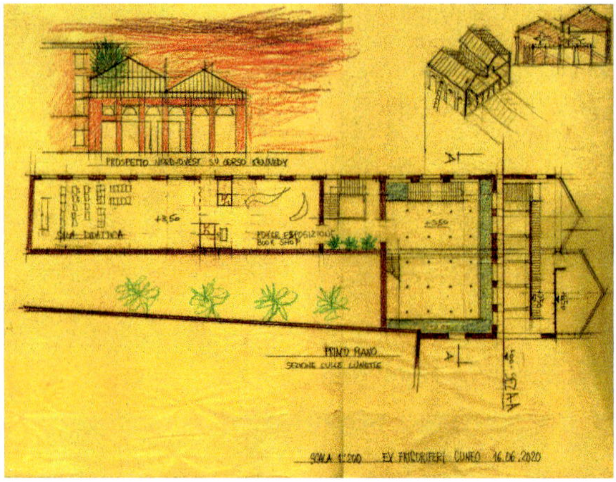

Fig. 1.28 Compatition Project, Rigeneration of ex military refrigerator, First floor project

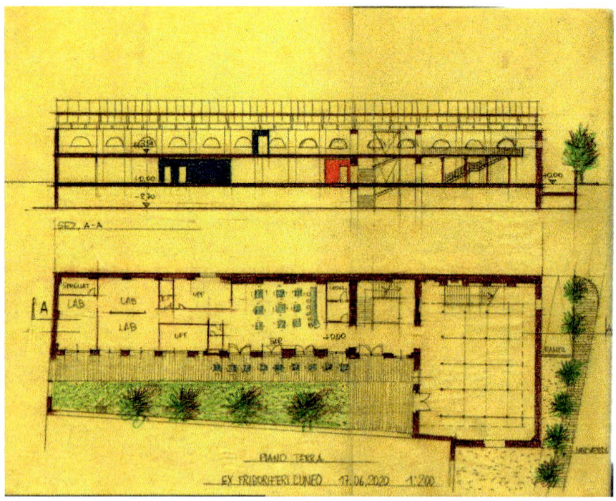

Fig. 1.29 Compatition Project, Rigeneration of ex military refrigerator, Ground floor project and section

Fig. 1.30 Compatition Project, Rigeneration of ex military refrigerator, Axonometric view

Fig. 1.31 Compatition Project, Rigeneration of ex military refrigerator, View of bookshop

Fig. 1.32 Compatition Project, Rigeneration of ex military refrigerator, View of the exhibition spaces

Fig. 1.33 Compatition Project, Rigeneration of ex military refrigerator, Exterior view, oil painting of Angelo Ariti

adopted. These are forms of expression capable of assigning a *meaning* to the new appearance attributed to the leftover, which goes beyond the simple re-purposing of space and is representative of the memory content it contains.

The figurative language offers the project the chance to free itself from functionalism and stereotypes.

It promotes the search for innovative answers, not taken for granted, to the problem of the regeneration of *leftovers*, as a priori, an argument not yet drawn from the experience and awareness of *leftovers*, but deduced from *reason*, from the meaning assigned to it, starting from research work prior to design. As far as tools and techniques are concerned, it entails the use of low-cost, light, soft, *hermaphrodite* materials—to use Mendini's term—even waste materials, and solutions that can be implemented quickly, that are reversible, temporary, if necessary even "seasonal", i.e. aimed at using space for limited periods of time. In terms of grammar, it provides for the use of measures that belong to the tradition of the project, in architecture, design, art, music, cinema and literature: from Mendini's change of scale, to reiteration, Varini's anamorphosis, perspective illusionism, the editing of Ėjzenštejn, the unbalancing actions of La Pietra, De Lucchi's *mise en scene*, the conceptual design of Guixè or Starck's perturbing design, to dissonance, to Stocchi's use of "magic" and enchantment, to the *Cut-up* of Burroughs, Castiglioni's *ready-made*, Sottsass's use of metaphor, the homoeopathic design of Bouroullec, and the erasure of Isgrò.

These are just a few examples of a "syllabary" necessary to form the words that make up the text.

A syllabary still to be created, but whose aim is to remove the design of the unfinished from the danger of simplifications that lead to chasing after the commonplace, also on the theme of contemporary living, or that lead to the delusions of the project participated in, an orphan of design. The *design of the unfinished* is unobtrusive and does not blink. It takes on the task of suggesting a new aesthetic code. It is a style of thought, aimed at creating shelters for the nomads of the third millennium. It renounces the polished image to give shape to the unfinished, the unprecedented and the unthought.

Appendix

Luciano Crespi, Competition project of new Student House, into former textile factory, Verrès, Italy, 2020

Consultant: ing. Paolo Saluzzi. Collaborators: Francesco Antonelli, Lucia Ratti, Mirco Sturlese

Honorable mention for the ability to evoke stories of people and memories of buildings

Fig. 1.34 Former textile industrial factory. Photo of Lucia Ratti

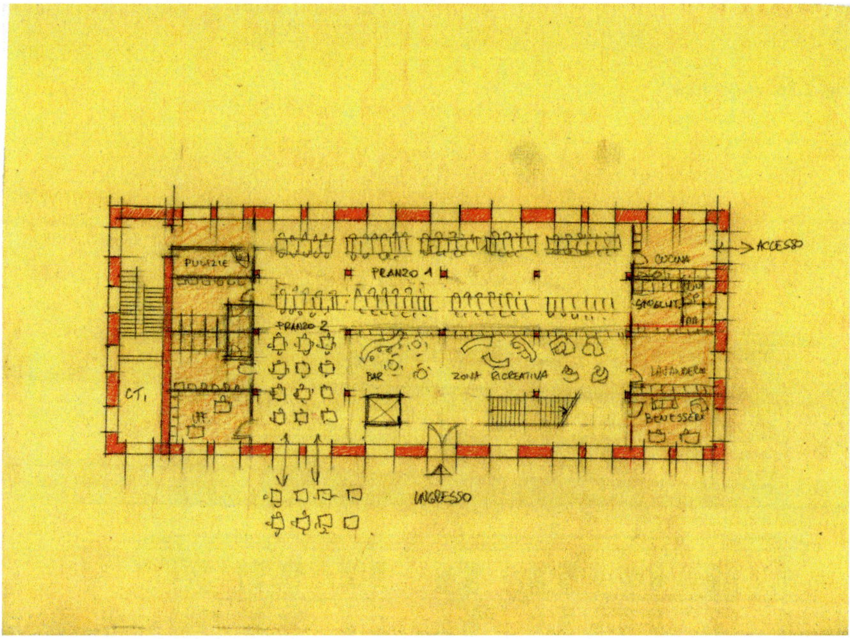

Fig. 1.35 Ground floor

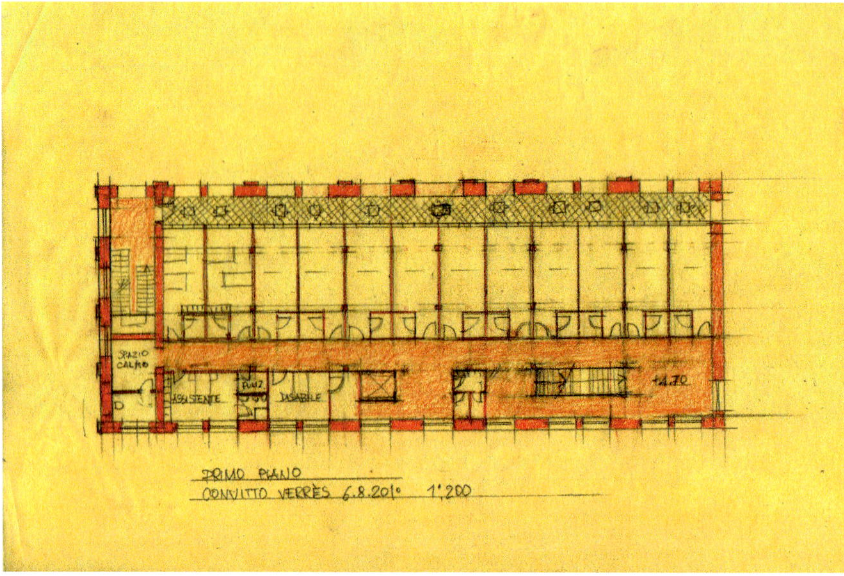

Fig. 1.36 First floor

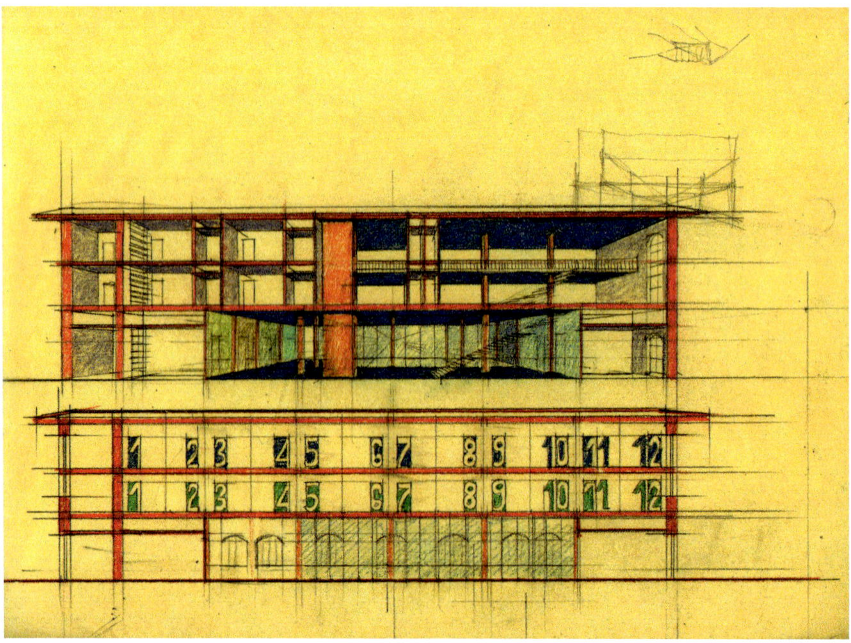

Fig. 1.37 Cutaway perspective and section

References

Argan GC (2005) Scultore per vocazione. In*:* *Michelangelo scultore*, serie "I classici
Bachelard G (1957) La poétique de l'espace. Les Presses universitaires de France, Paris
Benjamin W (2002) I passages di Parigi. Einaudi, Torino
Berger J, Mohr J (2017) Il settimo uomo. Contrasto, Milano
Boeri S (2011) L'anticittà. Laterza, Roma
Bourriaud N (2014) Il radicante. Postmedia Books, Milano
Brandi C (1963) Teoria del restauro. Einaudi, Torino
Branzi A (2016) Sull'autostrada per Karnac. Riflessioni sulla metropoli primitiva. Postmedia Books, Milano
Brusatin M (2000) Arte dell'oblio. Einaudi, Torino
Campagnoli G (2014) Riusiamo l'Italia. Da spazi vuoti a start-up culturali e sociali. Il Sole 24 Ore, Milano
Caniggia G, Maffei GL (1982) Composizione architettonica e tipologia edilizia. Lettura dell'edilizia di base. Saggi Marsilio, Venezia
Capanni F (2019) Dio non abita più qui? Dismissioni di luoghi di culto e gestione integrata di beni culturali ecclesiastici. Artemide, Roma
Clément G (2005) Manifesto del terzo paesaggio. Quodlibet, Macerata
Dagnino A (1996) Nuovi nomadi. Pionieri della mutazione, culture evolutive, nuove professioni. Castelvecchi, Roma
Dagnino A (2016) Nomadi transculturali, caravanserragli urbani e spazi pubblici di quartiere. In: Barbara A, Ceresoli J, Chiodo S (eds) *Interni inclusivi. Dialoghi trasversali.* Politecnica, Sant'Arcangelo di Romagna
dell'arte" n. 64. Milano: Skyra/Rizzoli
Galimberti U (2000) Orme del sacro. Il cristianesimo e la desacralizzazione del sacro. Feltrinelli, Milano
Lingiardi V (2017) Mindscapes. Psiche nel paesaggio. Raffaello Cortina Editore, Milano
Maffesoli M (1997) Du nomadisme. Vagabondage initiatique. Le Livre de Poche, Paris
Martí Arís C (1993) Le variazioni dell'identità. Il tipo in architettura. CittàStudi, Milano
Mendini A (1982) Il nuovo soggiorno. Domus, n 630
Sassen S (2006) Why cities matter (Perché le città sono importanti, in Città. Architettura e società. Marsilio/Biennale di Venezia, Venezia
Sennet R (2006) Il declino dell'uomo pubblico. Bruno Mondadori, Milano

Chapter 2
The State of the Art Between Needs and Desires. Design of the Unfinished as a New Perspective of Intervention on Existing Building

Fiamma Colette Invernizzi

Abstract Stop. Listen. Care. These are the three actions that make the design of the unfinished a valid approach on contemporary interiors: move away from the speed of a world that does not pay attention to details, history and memory of places in order to go deeper into a dimension of observation and attention to what exists and surrounds us, in the beauty of contemporary ruins and crumbling walls capable of telling old stories. As Franco Arminio wrote in one of his poems, "More than the year of growth, we need the year of attention. (…) Today being revolutionary means taking away more than adding, slowing down more than accelerating, it means giving value to silence, light, fragility, sweetness". Cracks that line surfaces—like buildings' wrinkles—and small design flaws become worthwhile opportunities to explore an increasingly widespread design in Europe and around the World, in a succession of unique microcosms that show the different ways of dwelling the leftovers and living contemporaneity.

Keywords Leftovers · Interior design · Perspective · Challenge · State of the art

2.1 Introduction

Stop. Listen. Care. Three verbs, three actions with which to begin the perspective and the journey toward forgotten, severed spaces, segregated in the corners of cities that go unnoticed and in the shadow cones. Three verbs that indicate completely different actions than those in which we all grew up, from the early nineteenth century to date, from the spreading of the first industrial revolution, through economic booms and overproduction, to the present day. The Covid-19 pandemic certainly forced us to stay home, but we are not sure yet that, after the vaccine, there can be a real turnaround. The paradigm shift is more necessary than ever, to reverse a course that speeds up and increases the climate emergency.

F. C. Invernizzi (✉)
Dipartimento di Design, Politecnico di Milano, Milano, Italy
e-mail: fiammacolette.invernizzi@polimi.it

L. Crespi (ed.), *Design of the Unfinished*, The Urban Book Series,
https://doi.org/10.1007/978-3-030-73457-2_2

Buried by tons of concrete, the construction mechanism never stopped and on the contrary, it underwent a powerful acceleration from the sixties to the 2000s, manifesting a sadly anthropocentric current of thought—selfishly anthropocentric.

Stefano Boeri wrote in the editorial of Abitare issue 480, 2008, "When thinking about their work, it is increasingly common that those fields who deal with the planning and transformation of lived space also have a strong ethical and philosophical edge to their outlook. We are pushed today towards thinking about the world in terms of the survival of humanity, thanks to an ongoing environmental crisis, unstoppable population expansion, the widespread urbanisation of the planet and the destruction of our natural resources (in terms of animals and nature)—in other words thanks to a kind of species suicide. It is no longer enough simply to look after the principles, values and the needs of our species, we also need to locate these within a wider vision of the future of the planet".

Twelve years later, the pandemic in progress has accelerated well-known and already underway trends, showing the fragility of a species—the human one—which until now has always believed itself invincible. Today is the time to take care. Paul Hawken says the same, when in Blessed Unrest he states with an elegant body-world parallelism that "We can, however, protect, nourish, listen and take care of our bodies with food, sleep, prayer, friendship, laughter and exercise. This is all the planet asks of us: allies, rest, nourishment, respect, celebration, collaboration and commitment". The action, then, is proposed to us by the youngest and most well-known activist in the world, the Swedish Greta Thunberg, when at 16 she pronounced in an international TEDx conference "And yes, we do need hope. Of course, we do. But the one thing we need more than hope is action. Once we start to act, hope is everything. So instead of looking for hope, look for action. Then, and only then, hope will come".

And here research and design come into play, at the service of complexity. Here comes into play the beauty of discovering the facets of the existing building and the architectures that come into this precious book considered to be the "unfinished". The interior design culture seems to be equipped, by history and attitude, to play a leading role through the adoption of an original design concept, different than the one of the building restoration or renovation, but also including the furnishings as well as is usually understood. The challenge is to experiment with interventions that can give these spaces new possibilities of use, by employing staging equipment that is temporary and reversible, provided that they are consistent with the nature and soul of the place, in order to facilitate its reintegration into the living social tissue and the enhancement of their symbolic content.

2.2 Seven Hints

Exploring the design of the unfinished and the aesthetics of the leftovers, during the years of my PhD, allowed me to get in touch with numerous design philosophies similar to those presented in this book, translated into architectural and design language. I take this opportunity here to tell the state of the art by following seven ideas and definitions—sometimes irreverent, romantic or proactive—that I think go well with the sense of design of the unfinished, with its beauty and complexity. Some thoughts are irreverent, others deliberately provocative, others linked to the passion I feel about this theme, to the sensitivity and beauty intercepted in these years of study. All are to be understood as food for thought for a broader thought and as a starting point for the creation of an autonomous and valuable reality.

(Almost) every idea is accompanied by design references and related QR codes, which refer to the virtual pages where you can find numerous project images. This choice is the result of wanting to activate the reader in the search for beauty.

2.3 Is the Design of the Unfinished a Desire?

To understand this question, the reflection must be made starting from a step backwards, from the difference that exists between need and desire. The first, in fact, which unites humans to beasts, brings us back to a sphere of basic necessities, of shelter from bad weather and dangers, from the search for food and water, to that of reproduction for the protection of the species. But the moment the human being—or actually his hairier and less developed ancestor—started walking on two legs (or legs) instead of continuing to crawl, a gigantic change of perspective took place—an exponential widening of the visual horizon. If walking on all fours paws allows a living being to look from the self just up to the horizon (mostly, and without effort, to control the surrounding territory composed of prey and predators), the bipedal gait makes possible a gesture of unexpected power: look at the stars. And so, our dear ancestor puts his ego—and his need—in a relationship for the first time in history with the new dimension of the sky and with it the infinite and the unattainable. The traces of this movement and the astonishment of our ancestor toward the starry sky have remained imprinted in the Italian language—and also in the English verb "desire"—thanks to the origin of the word from the Latin term "de-sidereum", meaning coming from the stars.

Applying this anthropological and awareness step to the design of the unfinished represents a first starting point and activation for a broader reflection. Is the design of the unfinished really something so distant—like the stars—to aspire to but never reach? No. And the evidence is there. It is certainly not a simple subject—because it is not as pure as mathematics—nor an exact science or a language with a linear construct, as it draws its roots in multiple teachings and concepts, including beauty, memory, history, anthropology, aesthetics, living, architecture and design.

What is certain is that the **Design of the unfinished is for the brave**. Shy and not too cheeky, he is less shouted at in sector publications, sprinkled with green architecture and glass skyscrapers, large real estate projects or futuristic infrastructures. This happens in a world of media giants, in which gigantic architectures invade gigantic international markets. No criticism, we are just talking about different dimensions, of realities that work at different scales. The former move large amounts of capital, the latter have a domestic measure, which rolls up its sleeves and finds an innovative solution even where there seems to be none. A clear example of this approach—and this courage—is Carles Oliver, a young and brash architect from Mallorca who made him the concept of "NOT TO DO, as the best way to do". Cultured and sensitive to the environmental situation, it demonstrates an approach in contrast with what is usually taught in academic faculties (and yes, we will return to this point later). As if to provoke the world of architectural composition, Carles Oliver proposes a method of intervention made up of attention to detail and courageous gestures—such as widening a hammered opening and then leaving all its marks in plain sight—for a contemporary and alienating, shrewd and definitely inexpensive intervention. Now increasingly published, his approach is evident in his latest two projects: the St. Miquel 19 Refurbishment, and the Can Lliro Coffee-Concert Refurbishment, both in Mallorca. The signs of the previous exposed false ceilings, as well as the lack or partial replacement of fixtures, the reuse of poor or existing materials are the basis of his work and his philosophy: an approach not seen, not sufficiently studied, which is gaining ground and appreciated by users, as well as—very slowly—also by the public and the publishing and constructive sector.

Reference projects:

- *St. Miquel 19 Refurbishment, Mallorca, Spain, 2016, by Carles Oliver (left)*
- *Can Lliro Coffee-Concert Refurbishment, Mallorca, Spain, 2019, by Carles Oliver (right)*

On the other hand, the **Design of the unfinished is a way of thinking**. And perhaps also a way of being. A way of thinking that recalls the need for a way of communicating and teaching. Following Carles Oliver's "not to do", I also think of the great teaching of Filippo Grandi, Commissioner of the United Nations High Commissioner for Refugees, when he says that "Western civilization must learn to have less, in order to have more". The Design of the unfinished refers exactly to this philosophy, considering the new construction heritage as a useless surplus of concrete that does not bring added value. Not to be drastic or anarchic, certainly the newly built heritage can be fundamental but it certainly must first go through the question "is it really necessary?". The design of the unfinished, the reuse and attention to the existing are concepts that have not yet entered the dynamics of thought and university training courses with the size and power with which they should.Starting from people grown during the modernist economic booms, the most widespread thoughts in technical-architectural universities are still linked to the overbuilding of new architecture against freeing natural soil.

It is up to us—perhaps even us, the authors of this book—to continue in the struggle and in the will to share and teach the techniques and thinking of what I dream of becoming a discipline, a degree course. Teaching that in the world, in addition to *archistars*, there are sensitive designers such as Nauzet Rodriguez or the R-Zeros who in Mexico have produced spaces of the highest quality and design level, with few and careful interventions. Allowing younger students to get to know an alternative path not only to that of the design of the new but also to that of restoration in its purity, in a mixture of disciplines and ideas that is varied, complex and even more interesting for this reason. The attention to the material detail, to the existing surfaces and light, to the patina of time that inexorably rests on every floor, to the scent that remains, to the history that can be read in the floors, all this should enter into the university with the mastery of a new and ancient language, with the clarity of a sensitive need.

Reference projects:

- *CDLE Offices, Mexico City, Mexico, 2015, by R-Zero (left)*
- *Colonial House on 64th street, Merìda, Mexico, 2016, by Nauzet Rodrìguez Ruiz (right)*

For the Design of the unfinished sensitive souls are needed, as well as sustainable ideas. Sustainable in the proper sense, from the dictionary definition, in which sustainability is the characteristic of a process or state that can be maintained at a certain level indefinitely. Thus, it becomes clear how, between sensitivity and sustainability, the concept of "taking care" of a reality, a space and a process, both cognitive and planning, fits. Care, as well as sensitivity and sustainability, requires a much longer time than that of simple construction, given that they are the result of material and imperceptible stratifications. Both the Voice of coffee, in Kobe, and the Factory

Life, in Belgium, are perfect examples of this. In the first case, Japanese rigor, cleanliness and simple elegance accompany refined design choices, in the second case the distribution of spaces, their flexibility and modular arrangement are emblematic of practical and effective thinking in the field of sustainability. Very different from each other, in size and aesthetics, both projects tell how the existing spatial character and the discovery of the materials already present on site are opportunities for reflection for a search for effective alternative and contemporary solutions, so that the life of the spaces can "be kept at a certain level indefinitely".

The Voice of coffee by Kobe takes advantage of every existing material detail as an added value, complying with the pre-existing elements, the missing false ceilings, the holes in the walls of the previous plants and the peeling surfaces, combining materials such as wood and glass, with very pure shapes and volumes, underlining their difference and highlighting their ability to coexist. Factory Life, on the other hand, adopts the concept of sustainability, through the shrewd choice of reducing the spaces used—within a much larger environment—to single and autonomous volumes, equipped with variable heating according to need. Pure volumes immersed in a semi-cold environment completely autonomous for livability and heating generate additional shells for domestic environments. Minimum energy expenditure, in order to join forces between cold (external), semi-heated (internal connection) and heated (autonomous domestic volumes) environments.

Reference projects:

- *Voice of coffee, Kobe, Japan, 2018, by Yusuke Seki*
- *Factory Life, Waarschoot, Belgium, 2012, by Julie D'Aubioul*

However, the age-old question that affects the **Design of the unfinished**—which we cannot fail to talk about—is the underlying risk that it will turn into an "**à la mode**" gesture, photogenic and captivating for certain luxury sector magazines, in which finishes and design choices fall into a dimension that is not that of respect for memory or care for the place. There is a risk—and in some respects, it is also legitimate not to demonize the architecture of luxury as too often happens—and it should only be observed as a phenomenon and as one of the facets that this approach can take. The border is blurred and thin and why demonize one approach to another? If the luxury applied to existing buildings can be a dynamic so that they are not forgotten or do not end up in ruins, then it is still an approach that can be updated and from which to benefit. If, on the other hand, it is a means of revolutionizing a space, through unsustainable material choices and approaches (even economically, if not for an exclusive audience), then the phenomenon must be analyzed more carefully.

Difficulty or not, skepticism or not, the design of the unfinished can always be defined as a gesture of love. A gesture that—as we have already seen in part—mixes care, listening and attention. An approach that requires observation and study, time and dedication. A dynamic that brings with it the power of the past and history, its aesthetic and image aspects, its amount of memory and *transdisciplinarity*, in a path in which designers are confronted with questions such as "what kind of relationship with the existing building (its aesthetic, functional, material, spatial… aspects) do I choose?", or "what role can the signs of the past and deterioration play in the renovation project?".

"Not designing is the challenge", says, for example, the designer Nauzet Rodriguez, "and that the project itself goes unnoticed, invisible, making the result suitable for the building, provoking the idea that what we see today has always been there and the only thing that has done has been aging. There are some recurring elements", explains Nauzet Rodrìguez Ruiz, "in this type of design action, which could be considered in some way as a style line, whether it is the preservation of aged finishes, visual installations, patched surfaces that preserve original fragments". Harmonize, therefore, maintaining the spirit and flavor of the past as fundamental ingredients of an agreement composed between past and present. "Remnants of a building's past that are uncovered and celebrated have the potential to instil into people a sense of appreciation for the people who came before them in a very tangible way", says Neumann Monson—from Neumann Monson Architects. "We often have an intense focus on the now and the new and the next things. We lose an appreciation of process, legacy and history. Material things too easily become dispensable and replaceable. New, re-use aesthetic should be based on preserving and celebrating the historical character of a place while allowing new ideas and Architecture to grow and thrive. In order to do this well, I believe it's critical to understand the historical value of an existing Architectural aesthetic in order to discover its potential to be redefined and added to".

Gestures of extreme attention to the existing architectures, which tell the need not to follow the capitalist and instrumental economy of soil consumption and exploitation of world resources. Or, at the very least, to know that an alternative route is possible and extremely valid.

2.4 Conclusion

I would like to conclude, therefore, by saying that the **Design of the unfinished is NECESSARY**. And I would like to describe this point through the words of one of the greatest intellectuals and poets of our time, a committed "landscape scientist" who, in moments of crisis and difficulty, acts as a shrewd mirror of the reality that surrounds us. He tells of a critical moment—either because of the pandemic, the social situation and the economic uncertainties—by writing publicly, on his social channels—Thursday, December 17, 2020—about the betrayal of intellectuals and politics toward generosity.

"The intellectuals betrayed us and politics betrayed us. They were left speechless when it was needed most. Words are not what you decide to say or not to say, they are those that come out when you forget about yourself and your conveniences, the tongue is heard when it escapes from your hand, when it gushes directly from the body, when that discursive film skips, that academic envelope that serves to say I know things, rather than to really say things. They betrayed us, but the world is not finished, the generous are making up for the desertion of the traitors. And the generous are in the South and in the North, on the right and in what was once called the left, they are between the poor and the rich. The generous have no film, the gift gets out of hand, it reaches you concrete, tangible. The traitors did not do it on purpose, perhaps they do not gain anything from their betrayal, it is that they are like that and what they are is no longer enough, they are people who do not have the keys to enter places, do not have the eyes to see people and the world we have before our eyes. The traitors had formed in a world that proceeded by abstractions, by great theoretical draperies that seem to envelop the world and understand it. But at some point, the world decided that it did not want to be understood anymore, this is the virus, before a disease is an element that adds up and blows everything up. Now we are in this very difficult passage and we must understand, this we must do. In this passage the very ones who by definition were responsible for guiding and interpreting the world were found unprepared, but nothing is finished, the sunrises and sunsets continue, the tremors of desire and the care of those who care and the hopes of those who want to be cared for, there are workers and there are peasants, there are those who work and in addition to working produces a thought, a vision that is lacking in the deputies. This situation changes from day to day, has no definable characteristics, on the one hand the feeling of closeness seems to collapse, but then you realize that it reforms where you least expect it. It seems that we are dying under the black asphalt of depression and then you feel that the asphalt splits, tufts of passion, of joy appear. Perhaps this is the only thing we can do now, watch over the cracks, only from there can something come out".

Watching around the cracks is then the only action to be taken, to see the wonders blossom from the existing, to treat them with generosity, to look at them with care and attention so that they mix with the existing beauty and give meaning to this generation of Prime rubble, bruised but not demoralized. Kindness in a new design, in a new teaching, in a new generation of designers committed to safeguarding the existing in its fragility, considering the Design of the unfinished as the first step for a new discipline.

Bibliography

Abitare AAVV (2008) rivista di architettura, arredamento e design, n°480

Anzani A, Guglielmi E (A c. Di) (2017) Memoria bellezza e transdisciplinarità: Riflessioni sull'attualità di Roberto Pane. Maggioli editore, Santarcangelo di Romagna (RN)

Appiano Caprettini A (1999) Estetica del rottame: Consumo del mito e miti del consumo nell'arte. Meltemi, Roma

Arminio F (2017) Cedi la strada agli alberi. Chiarelettere, Milano

Astrologo M (2014) Vite di scarto. Laterza, Roma; Bari

Augé M (2004) Rovine e macerie. Il senso del tempo. Bollati Boringhieri, Torino

Bailly J-C (2016) La frase urbana. Bollati Boringhieri, Torino

Baudrillard J (2010) La società dei consumi: I suoi miti e le sue strutture. Il mulino, Bologna

Baudrillard J, Codeluppi V (1987) Il sogno della merce. Lupetti, Milano

Bauman Z (2007) Lavoro, consumismo e nuove povertà. Città aperta, Troina (EN)

Bauman Z, Cupellaro M (2015) Consumo, dunque sono. Editori Laterza, Roma; Bari

Calvino I (2014) Le città invisibili (ristampa). A. Mondadori, Milano

Clément G, De Pieri F (2014) Manifesto del terzo paesaggio. Quodlibet, Macerata

Crespi L (2013) Da spazio nasce spazio: L'interior design nella trasformazione degli ambienti contemporanei. Postmedia books, Milano

Crespi L (2018) Manifesto del design del non-finito. Postmedia books, Milano

Crisman P (2007) From industry to culture: leftovers, time and material transformation in four contemporary museums. J Architect 12(4):405–421

Dagnino A (1996) I nuovi nomadi: Pionieri della mutazione, culture evolutive, nuove professioni. Castelvecchi, Roma

Finiguerra D (2014) 8 mq al secondo: Salvare l'Italia dall'asfalto e dal cemento. EMI, Bologna

Friedman Y (2009) L'architettura di sopravvivenza: Una filosofia della povertà. Bollati Boringhieri, Torino

Hawken P (2008) Blessed unrest: How the largest social movement in history is restoring grace, justice, and beauty to the world. Penguin Books, New York

Hebel D (A c. Di) (2014) Building from waste: recovered materials in architecture and construction. Birkhäuser, Basel; Boston

Hillman J (2004) L'anima dei luoghi. Rizzoli, Milano

Latouche S (2012) Limite. Bollati Boringhieri, Torino

Latouche S, Schianchi M (2015) La scommessa della decrescita. Feltrinelli, Milano. Le case dell'uomo: Abitare il mondo (2016) Torino: Utet

Lynch K, Ceccarelli P (2008) L'immagine della città. Marsilio, Venezia

Maroldi F (2014) (S)punti: Frammenti testuali (Collana FLEC)

Mumford L, Rosso M, Scrivano P (2007) La cultura delle città. Einaudi, Torino

Neuwirth R (2007) Città ombra: Viaggio nelle periferie del mondo. Fusi orari, Roma

Sassen S (2006) Perché le città sono importanti. Città. Architettura e società, Marsilio edizione

Sassen S (1999) Migranti, coloni, rifugiati: dall'emigrazione di massa alla fortezza. Europa. Feltrinelli, Milano

Sassen S, e Appiah A (1998) Globalization and its discontents: essays on the new mobility of people and money. The New Press, New York, NY

Scanlan J (2005) On Garbage. Repr. Reaktion Books, London

Simmel G, Jedlowski P (2012) Le metropoli e la vita dello spirito. Armando, Roma

Speroni F (2002) La rovina in scena: per un'estetica della comunicazione. Nautilus 7. Meltemi, Roma

Thunberg G (2019) The disarming case to act right now on climate change [Video]. TEDx Conferences
Vidler A (2006) Il perturbante dell'architettura: saggi sul disagio nell'età contemporanea. Einaudi, Torino
Virilio P (1992) Estetica della scomparsa. Liguori
Zevi B (2009) Saper vedere l'architettura: saggio sull'interpretazione spaziale dell'architettura. Einaudi, Torino

Chapter 3
Regenerate Urban Leftovers

Barbara Di Prete

Abstract In the contemporary city, there are countless public and private spaces that are in a temporal and functional "limbo", often facing an uncertain future; many current examples, however, show how these "silent" products can be reintegrated into the social fabric, investing relatively low resources and triggering virtuous processes to enhance the territory with a view to economic sustainability, functional reversibility and cultural identity. Indeed, their metaphysical nature is well suited to make way to new design "writings". These "leftovers" of the architectural heritage turned into cultural and creative hubs, often incubators of new production models, are also opportunities to experiment with new widespread refunctionalisation processes that, from the interiors, prove to be also capable of upgrading the city's public spaces. These actions have encouraged integrated public-private partnerships, often leveraging art and culture to activate the local spontaneous resources and, on longer horizons, to change the urban image (and imaginary).

Keywords Leftovers · Cultural and creative hubs · Collaborative experimentation · Multicentric peripherality

3.1 Introduction

This chapter aims at investigating an increasingly widespread phenomenon of refunctionalisation that not only starts from interiors, but also affects the surrounding urban environment and, sometimes, even the urban layout, reading the transformation that has involved several buildings once used for production (but not only) purposes and today rethought as cultural and creative hubs, places of exhibition and research.

The chapter specifically analyses some pilot cases of the Italian and European context that offer occasions of reuse and have had a deep impact on the physical organisation of the city, as well as on its symbolic and relational structure. Both

---

B. Di Prete (✉)
Dipartimento di Design, Politecnico di Milano, Milano, Italy
e-mail: barbara.diprete@polimi.it

L. Crespi (ed.), *Design of the Unfinished*, The Urban Book Series,
https://doi.org/10.1007/978-3-030-73457-2_3

the urban image and imaginary are thus changing in the light of a new "multi-centric peripherality" and of a "collaborative experimentation" between public and private, keys to reading that could be taken as categories for interpreting the ongoing revolution.

The interest for these abandoned spaces is shown by a variety of increasing initiatives around them that also entail a heterogeneity of sometimes antithetical approaches.

The 2016 publication of the volume *Sguardi tra i residui. I luoghi dell'abbandono tra residui, utopie ed eterotopie* (Giulia Dal Borgo et al. 2016) has then been followed by many more books on the issue. Among these, *Riusiamo l'Italia* (Campagnoli 2014) addressed the topic in a systematic way supporting the reading with analytical data, from which an interactive map was created (http://mappa.riusiamolitalia.it) that can be implemented by users individually to put demand and offer of abandoned places in contact "from the bottom".

Furthermore, the so-called Urbex (abbreviation for "Urban Exploration") phenomenon is particularly popular, consisting in an urban exploration aimed at discovering, capturing and thus restoring the charm of abandoned, often ruined buildings. Such practice was also supported and spread by the media through successful TV shows (*Urban Explorers, Cities of the Underworld and MTV's Fear*). The motto of those nurturing this passion is, however, quite telling: "leave only footprints, take only emotions" (Nestor 2007); it implies an exclusively contemplative approach, without any modifying intent. On the other hand, those dealing with "leftovers" today—at administrative as well as design level—consider them much more than just testimonies; the shift from contemplative dimension to concrete action characterises the current policies, aimed at filling these "gaps" in terms of functional and symbolic recovery, of temporary reactivation for events, as well as for temporary exhibition setups.

This issue is particularly important in Italy, as in the last 70 years the territory has ont only increased its built environment by 400% but has also seen an exponential increase in the number of abandoned spaces; to date, the unused or underused properties are thought to be over six million (Maccaferri 2016). The relevance of the problem and the potential of the adaptive reuse of this huge building complex is therefore apparent.

3.2 A Collaborative Experimentation Between Public and Private

It is interesting to highlight the positive co-management experiences or the establishment of collaborative agreements for the reactivation of the so called "urban leftovers" starting from integrated public-private-third sector partnerships that, on the one hand, concern the property redevelopment phase and subsequently its refunctionalisation. The latest policy guidelines establish that the administration guarantees the fee-free or highly facilitated use of the public building versus a medium-long management time commitment (usually at least 10 years, but 30-year agreements are also not uncommon). In return, the private body involved must bear the costs for all the necessary works for bringing the building back into operation. In this way "the process responds to a plurality of interests: the need to experiment with innovative management procedures and flexible intervention policies, the need to trigger an active involvement of the local community, the opportunity to identify a vision capable of transforming buildings of collective interest into drivers of area development" (Di Prete in Anzani 2020, p. 185).

In this perspective, in the Italian case the initiative aimed at turning abandoned buildings into art ateliers or cultural centres is worth mentioning, thanks to a government call insisting on a census of abandoned properties (Ministerial Decree 18th December 2018); the call—promoted by the Ministry of cultural heritage and activities—envisaged a 10-year concession of these buildings to groups of artists for the low price of 150 euros per month, encouraging the most interdisciplinary projects willing to share the spaces with artists coming from different areas, promoting open events and fruitful relationships with local stakeholders, at the end capable of guaranteeing financial sustainability and a connection with the cultural fabric of the territory.

The young artists (preferably under 35) must bear the full costs of ordinary maintenance and partially the ones of extraordinary maintenance, being also allowed to start specific sponsorship agreements with private institutions (Mammarella 2019).

Following the national model, many local initiatives have also been carried out with similar purposes. The Milanese one is certainly a pilot case-study, which has been experimenting for some years now with this public-private-third sector partnership starting from calls for applications (upon specific project proposal) or from temporary concession agreements for events.

The latest is the one just promoted by the Municipality of Milan to give long-term (30 years) free concessions (or subsidised rents) on 1,200 public housing areas, mainly located in peripheral districts, with the aim of relaunching the economy and "recreating the community" (Bonomi in Dazzi 2020).

In this regard, the cases of Nuovo Armenia, Asnada and Design Open Spaces (DOS) are emblematic.

Nuovo Armenia—which echoes the name of the historic Milan film factory—and Asnada are cultural associations working to foster multiethnic integration that has recently taken a common challenge, choosing to redevelop a historic farmhouse of Dergano, in Via Livigno no. 9, owned by the Municipality of Milan. The farmhouse was given for 30 years to the above associations in 2016 through a public call for the redevelopment of abandoned buildings: after a long restoration and space rearrangement work, which also involved the various local communities, today the building has become "a creative hub for urban regeneration, a place of action and inclusion where creative production processes are activated with cultural and social service functions" (www.cascina9.polimi.it).

These dynamics of inclusion and engagement of the local players often characterise contemporary redevelopment processes, whose original purpose is exactly to enhance local resources. In this way these "public assets" really become "common assets", as able to welcome and engage into these activities also the people residing in the district, triggering community animation pathways and participatory processes.

The model proposed by DOS—Design Open Spaces, explaining its mission in its articles of association, is different: DOS is "an organisation based on the synergetic collaboration between public institutions and private bodies that with their work lead to reopen forgotten spaces, working with design practices to make them available again for the community" (https://designopenspaces.it/manifesto). DOS has been working since 2019 to act as a widespread design district, specifically using the event of the Milan Design Week to reactivate "numb places" owned by the municipality, scattered in various areas of Milan, sometimes even in central areas. The aim, through the collaboration between public administration and design companies, is to make some properties available to turn them from holes without identity into new exhibition centres of Fuorisalone. The management model provides that the companies and designers taking part in the project (without having to apply for public calls, thanks to the pre-signed agreement between DOS and the Milan Municipality) can use these spaces at reduced prices for the whole duration of the event, committing to regenerate them to then open them again to the public. Being companies working in the field, their setup and interior restoration works are often sustainable for them, surely cheaper than the investment they should bear for a same-size space in central districts of the city. Likewise, for the Municipality this is a virtuous action that increases exhibition opportunities, tackling building degradation and leaving the administration the possibility to use the property during the year. The public-private collaboration model, thanks to the mediation of DOS that works here as a "dialogue and action facilitator", appears particularly effective and could represent a repeatable best practice.

Finally, it is interesting to highlight the design approach pursued in many of these cases, which can be attributed to what Luciano Crespi symbolically defined "Design of the unfinished", a design capable of "dealing with the issue of temporariness and precariousness, even in terms of aesthetic practices" (Crespi 2018, p. 55). A design that does not erase, does not eclipse, does not impose itself but "listens" to the place and makes it a scenery.

Caselli di Piazza Principessa Clotilde re-opening: before and after. Project by DOS—Design Open'Spaces emiliolonardodesign e Re.Rurban, with the collaboration of Falegnameria Curioni, Milan, 2019

Many are the examples of refunctionalisation of even large areas with a strong urban impact on the whole Italian territory, also in non-metropolitan cities. Important works, for example, have been carried out in Ferrara.

Here Consorzio Wunderkammer currently occupies the former Magazzini Generali, a building on the city's river dock that was renovated by the Municipality with regional funds and then granted with a public call to local associations that transformed it into a cultural and social innovation centre, with several active citizenship activities and design of local practices (Basso Profilo), intercultural promotion (Encanto), sport and culture aimed at enhancing the river area (Fiumana). With their countless activities, festivals and training courses, today Magazzini Generali "are a regeneration facility on a redeveloped area, interface between historic centre and river" (Fusari 2017, pp. 72–73). In agreement with the Municipality, even the large area of the former distilleries AlcEste (220,000 m^2 of surface area) was an opportunity to test the format "Building Communities", developed by studio KCity starting from a shared design process that saw the involvement of various stakeholders. In its conclusion, the Municipality and the "project communities" implemented the resulting strategic scenario giving life to "Youth AlcEste", a young, creative and low-cost neighbourhood that also won the Urban Planning Award of Urbanpromo (Cottino and Domante 2017, p. 73).

Finally, again in the city based in Romagna, it is worth mentioning the experience of Factory Grisù, a consortium of creative brands that set itself up in the former Fire Brigade barracks, owned by the Province. In 2013, the association Grisù asked to manage the property, a long-abandoned asset of 4,000 m^2, to set there an incubator of innovative firms linked to the creative, digital and cultural world, also in this case selected through a call allowing the assignees to occupy the spaces and carry out all the necessary works to start their activities at their own expense, but without any rental fee (www.factorygrisu.it).

A similar collaborative initiative combining top-down policies with bottom-up processes able to engage administrations, stakeholders and local players that took place abroad is the one experimented many years ago in Brussels by the association Citymine(d), mediating between owners of temporarily empty buildings and artists/associations working in the cultural and creative field looking for a workspace. The project Precare (1999) included 14 interventions of allocation of spaces with variable-term contracts from 6 months to 5 years; the results here achieved show how the temporary reuse of buildings through these artistic-cultural initiatives can work as a credible alternative to the problem of squatting, of safety and social cohesion, enhancing at the same time architectural assets and increasing the value of the area.

Therefore, these "leftovers" of the architectural heritage are more and more "opportunities to experiment shared, inclusive and collaborative processes between public institutions and local actors and can then become best practices and good governance policies" (Di Prete in Anzani 2020, p. 188). Such approach "bets that the very activation of public interest services, designed as 'common' (that is, as services managed and maintained by communities of users) may act as a driving force for urban and property development processes of abandoned areas" (Cottino and Domante 2017, p. 73).

3.3 A Multicentric Peripherality: Art and Creativity as Drivers of Adaptive Reuse

Analysing the phenomenon of adaptive reuse of "urban leftovers", at least two factors emerge that can already be deduced from the previous paragraph: their often peripheral position, where the real estate interest is lower, and their frequent conversion into cultural and creative hubs respectful to the place's architectural vocation.

3.3.1 For an "Emerging" Periphery

The outskirts more than any other place meet the challenge of the contemporary age, which must go beyond the concept of "peripherality" as an attribute of "peripheralisation" (Pezzi and Urso 2016, p. 2); indeed, the first refers to a purely territorial eccentricity, while the second identifies a process of socio-economic marginalisation. The two concepts are often overlapping, yet this condition can be reversed activating the *emergencies* (literally, "things that emerge") typical of each territory, that is, the available resources that make value to the system. The "urban leftovers" that sprinkle the outskirts of Western cities can to all effects be considered latent, widespread, accessible resources with disruptive potential.

Such interpretation confirms the reading of Franco La Cecla, according to which it is only at the edge of metropolitan life that we can find a chance for regeneration (La Cecla 1988). Among others, also Renzo Piano has remarked on the strategic centrality (notwithstanding urban eccentricity) of these peripheral areas: "the periphery is always accompanied by an insulting adjective: far away, abandoned, sad; in reality, 80% of people in a city live in the periphery and it is there that the energy lies. [The peripheries] are full of life, they are places of the future, all it needs is for cities to grow to include, not exclude them" (www.renzopianog124.com).

Therefore, peripheries themselves, places for too long deprived of real planning as wrongly considered "lost cities", have become the privileged territory of experimentation for a widespread and "endemic" refunctionalisation capable of triggering new welcome, participation and sharing processes.

Among the several case-studies that we could mention, the main examples include the recent experience of Social factory (www.designdifferente.it), located in a former factory of heat sinks for railway lines in the Bovisa district, in Milan's outskirts. This is a particularly interesting case for the richness and variety of activities it proposes, from the paradigmatic ones (cultural events, fablabs and coworking) that are generally more successful in recovering "leftovers", to other, definitely more peculiar and innovative, activities that also raise people's interest in the district and contribute to increasing its value. In fact, Social factory often works with municipality 9 of the City of Milan (very recent is the installation of IlluminaMI repeated in two parts of the district); here are some of the activities it carries out: it hosts in its own premises the cultural association "La Repubblica del Design", promoting in the Bovisa-Dergano-Lancetti district the same-name Fuorisalone area, that is the first and only one not bound by real estate and economic dynamics. It builds networks between local players—institutional, academic, business and third sector ones—also providing supporting infrastructures for meetings and debates. It manages a coworking of creative firms where to experiment collaborations aimed at promoting technological innovation. It hosts one of the first design-oriented fablabs of Lombardy, IDEAS Bit Factory, proactive maker of projects on the territory and not only producer for third parties. It hosts a new showroom of furniture and decoration objects that inherited the former collection of "designers modern art" and of "anonymous design" of Blitz, making it an innovative exhibition model (a showroom where furniture and objects are not only elements to "admire", but become "protagonists" of the space often usable by the coworkers themselves). Finally, it promotes the brand "Design Differente" that also expresses the peculiar identity of this new hybrid space with a production, exhibition and social vocation. It is an open space with a strongly industrial style where no *camouflage* or mimesis works were done, nor is it winking to "the new": on the contrary, according to the already described approach of enhancement of the unfinished, the space was created with reversible, flexible and cheap, yet not less spectacular and cosy, setups able to interact with the pre-existing environment, with its physical and symbolic traces.

Social Factory, company headquarters located inside a former factory in Bovisa, in the Milanese suburbs, opening 2019.

Social Factory, company headquarters located inside a former factory in Bovisa, in the Milanese suburbs, opening 2019.

"illuminaMI", installationpromoted by Repubblica del Design and Astronove, realised by Social Factory and IDEAS Bit Factory in the public space of Bovisa, in the Milanese suburbs. Pic. by Davide Ostoni

3.3.2 *Towards Cultural and Creative Hubs*

The second factor concerns the privileged transformation of "leftovers" into creative hubs, in which artistic processes, cultural initiatives and performances become tools to trigger new collective rituals.

Indeed, these places, thanks to their often metaphysical atmosphere, not only become perfect sceneries for art performances, setups, exhibitions or installations, but are also sufficiently spacious and suitable for flexible layouts to become incubators of innovative startups.

The examples confirming this trend are many: some "former leftovers" are now renowned art exhibition and cultural production centres (from Hangar Bicocca to Fabbrica del Vapore, without forgetting the emblematic case of BASE, in the former Ansaldo spaces, today social and cultural incubator); others have converted their traditional production of tangible goods to intangible ones (former glassware shop Livellara, now location of Spirit de Milan, where you can find again the old mood of the city through its culinary and music traditions); some instead (former Milan slaughterhouse, now hosting Macao) "through practices of special occupation with the aim of liberating art making it usable for as diverse an audience as possible" (Di Prete in Anzani 2020, p. 184); finally, others had an unexpected recovery exploring new functional pathways (former soap factory in Lodi, transformed into an art laboratory, dance and circus school).

Last but not least, it's useful to remind the upcoming inauguration of ADI Design Museum Compasso d'Oro, a "narrative museum"—as the designers Migliore and Servetto defined it (Flavia Chiavaroli 2019)—located in a recovered industrial area of about 3,000 m^2 between Fabbrica del Vapore and Fondazione Feltrinelli.

Also, many are the important examples abroad, such as the Chocolate Factory in London, a chocolate factory that was turned into one of the most interesting "creativity hubs" of northern Europe (with ateliers for artists, craftsmen and creative talents, promoting activities linked to contemporary theatre, cinema and literature) and that contributed to developing the cultural district of Haringey/Wood Green, a previously run-down area in North London. Similar experimentations can be found in Holland and France; in the first case, the old harbour of the city has become a lively cultural centre hosting multidisciplinary festivals, shows, exhibitions and events in a post-industrial space with unmistakable features ('NDSM-werf, Amsterdam); in the second case, former military barracks on the Garonne hosts, among other activities, an area dedicated to urban sub-cultures, spaces for street artists, an open-air gallery and alternative commercial spaces (Darwin ecosysteme, Bordeaux).

In summation, it can be said that art—exhibited, participated, denunciation art, art as research lab—is more and more explicitly getting into cities' redevelopment processes and, exactly from its most "damaged" interiors contributes to "making the city what it is", rebuilding a complexity of spaces and relationships.

3.4 Conclusions

In view of the above, the increasing number of functional conversion processes with a creative purpose seems to be able to "activate spontaneous resources present in the places and become the driver of urban regeneration as well as social innovation. Increasingly widespread solutions for adaptive reuse of buildings, aimed at solving peripheral situations of disposal or physical degradation, give the market and the citizens renewed spaces in terms of their functions and social microcosm" (Di Prete in Anzani 2020, p. 188).

In fact, these are actions capable of engaging different communities, often proposing a broad time schedule and interconnections between multiple activities. In this view, abandoned areas also become social incubators, enhancing the cultural contaminations already characterising European cities: in this regard they can be considered a metaphor of the contemporary age, ideal places to welcome increasingly diverse communities, symbols of a globalised society where "the world has become everyone's backyard" (Rifkin 2011, p. 391).

Obviously, today these needs of social repossession and functional reactivation are more and more connected with the urge to save economic, landscape and territorial resources. From many perspectives, therefore, they identify one of the most promising experimentation fields on the built environment in view of the future.

Despite the difficulty reading such a broad and varied phenomenon, some peculiar features are recurring elements in all the above examples:

- on the one hand, creativity is not only seen as a practice of divergent and imaginative thinking, yet it becomes a lever for urban regeneration and social reactivation: a tool able to talk to large audiences with the language of art and innovation, to subvert an "atrophic" destiny at low prices and, frequently, with a considerable environment transformation.
- in particular, many of the cases selected propose firstly redevelopment and secondly management processes that are also economically virtuous, promoted

by the public in synergy with the private environment and the third sector: these "associations, non-profit organisations or creative companies, by creating bridges between public and private, are taking on the onerous task of redeveloping abandoned factories and converting them into exhibition and research sites for cultural or productive experimentation" (Di Prete, in Anzani 2020, p. 182).

- from our point of view, the design approach with which these abandoned buildings are given back to the social and urban life is key, to reactivate them and permanently save them from "oblivion", or to show that—with specific actions—they can temporarily get available again for the community. These buildings simply need setup, reversible strategies capable of enhancing the place and show its inherent potential without having to put in place more impactful construction or architectural works, to become again testimonies of a renovated, always worth-preserving story: "they can represent a strategic resource for the territory, not just being available to perform new functions, but also holding memories and human stories that would otherwise be lost" (Crespi et al. 2017, p. 1.473).
- finally, we highlight how these "reactivations" of abandoned artefacts become a concrete job opportunity for the owners of the many startups that most invest on such operations. In this regard, the "leftovers" must be considered as territories to explore for a virtuous reuse, also in a view to an employment impact and a financial return.

The most useful thing at this time in history—that sees increasing situations of abandonment, but also a renewed consciousness on the ways to fight it—would then be to "circulate" as many situations as possible to help the relaunch of the country-system starting from the enhancement of its already available resources. Obviously, the best situations are the almost ready-made abandonment or dismantling ones, that is, those where restoration costs and rehabilitation expenses are extremely low and, subsequently, where the "places colonisation" time is very short. As recently stressed by Roberto Tognetti in a conference held at Milan Politecnico (Tognetti 2020), "it is not a real estate transactions market, but, as opposed to that, a way to speed up value creation processes, that can and must be triggered even by temporary reuse activities, of creative reuse where *the values* of the content restore *the value* of the property and not the other way around".

References

Campagnoli G (2014) Riusiamo l'Italia. Da spazi vuoti a start-up culturali e sociali. Gruppo24Ore, Milan

Chiavaroli F (2019) Presentato l'ADI Design Museum Compasso d'Oro di Milano. Intervista a Ico Migliore (2019) In: *Artribune*, 31 Genuary 2019. https://www.artribune.com/progettazione/design/2019/01/adi-design-museum-compasso-oro-Milano-intervista-ico-migliore/. Accessed 11 Dec. 2020

Cottino P, Domante D (2017) I commons, volano di un nuovo sviluppo urbano. In: ECOSCIENZA, n. 5, Anno 2017, p 73

Crespi L (2018) Manifesto del design del non-finito. postmedia, Milan

Crespi L, Anzani A, Caramel C, Crippa D, Di Prete B, Lonardo E (2017) Designing remains. In: Amoruso G (ed) Putting tradition into practice: heritage, place and design, Proceedings del 5th INTBAU international annual event, Politecnico di Milano. Milan, 5–6 lug. Springer, Cham, pp 1473–1482

Dazzi Z (2020) Il Comune offre gratis ai cittadini gli spazi abbandonati nei quartieri. In: La Repubblica, 10 December 2020, https://milano.repubblica.it/cronaca/2012/09/29/news/comune_le_nuove_regole_per_gli_spazi_abbandonati-43494208/. Accessed 10 Dec. 2020

Di Prete B (2020) Dreams. The cultural and creative conversion of abandoned buildings as a driver of urban regeneration. In: Anzani A, (ed) Mind and places. Springer, Cham, pp 179–190

Fusari R (2017) Rigenerazione urbana, l'esperienza di Ferrara. In: ECOSCIENZA, n. 5, Anno 2017, pp 72–73

Giulia Dal Borgo A, Garda E, Marini A (eds) (2016) *Sguardi tra i residui. I luoghi dell'abbandono tra residui, utopie ed eterotopie.* Mimesis, Sesto San Giovanni

La Cecla F (1988) Perdersi. L'uomo senza ambiente, Laterza, Bari

Maccaferri A (2016) Spazi rigenerati con l'università del riuso. In: nòva, Sole24Ore. https://nova.ilsole24ore.com/progetti/spazi-rigenerati-con-luniversita-del-riuso/?refresh_ce=1. Accessed 10 Dec 2020

Mammarella P (2019). Progetti per trasformare edifici abbandonati in atelier d'arte, il Governo ci riprova. In: Edilportale. https://www.edilportale.com/news/2019/02/progettazione/progetti-per-trasformare-edifici-abbandonati-in-atelier-d-arte-il-governo-ci-riprova_68734_17.html. Accessed 9 Dec 2020

Nestor J (2007) The art of urban exploration. In: San Francisco Chronicle, 16 August 2007

Pezzi MG, Urso G (2016) Coping with peripherality: local resilience between policies and practices. Ital J Plan Practice VII(1):1–23

Piano R (2018) G124. Renzo Piano. Il gruppo di lavoro del senatore sulle periferie e la città che sarà. www.renzopianog124.com. Accessed 29 Oct 2018

Rifkin J (2011) La civiltà dell'empatia. Mondadori, Milan

Tognetti R (2020) Riusare il costruito tra innovazione e progetto, conference held in Milan, Politecnico di Milano, Scuola del Design, 17 November 2020

www.cascina9.polimi.it

www.designdifferente.it

www.designopenspaces.it/manifesto

www.factorygrisu.it

www.mappa.riusiamolitalia.it

www.riusiamolitalia.it

Chapter 4
Reusing Leftovers: Corporeity and Empathy of Places

Anna Anzani

Abstract Massive urbanization and exploitation of nature have produced unprecedented historical and natural transformations; this puts interior design ahead of the need of reusing existing urban spaces, often characterized by relevant historical architectural features. Reusing urban leftovers can facilitate place attachment, intensified by awareness of the place history: as shown by research in environmental aesthetics, historical places allow people to feel a sense of continuity, arouse curiosity and increase motivation to discover the forgotten past of places. Interesting experiments carried out by cognitive neurosciences in the architectural field have shown that the relationship between senses and art/architecture turns out to ground primarily on an emotional and multisensory experience. More than other arts, architecture generates multisensory impressions, which greatly involves our corporeity. Starting from these assumptions, leftover reuse design should place at its centre the people who live in places, overcoming formal and purely visual principles on which interior projects have often been founded.

Keywords Places · Perception · Memory · Corporeity

4.1 Introduction

Abuse of nature, space obsolescence and abandonment, and mass migration are deeply affecting our contemporary, questioning the condition of human coexistence. Understanding the complexity of our society, the revolution of working conditions and human relationships, the systemic connection between the planet and people's health requires a transdisciplinary perspective (Anzani 2020).

Promoting an intersection between interior design and other disciplines like psychology and ecology can help searching design approaches focused on emotional and multisensory experience, giving value not only to a functional, but also to an experiential use of places, with this term indicating spaces endowed with meaning,

A. Anzani (✉)
Department of Design, Politecnico di Milano, Via Durando 10, 20158 Milano, Italy
e-mail: anna.anzani@polimi.it

L. Crespi (ed.), *Design of the Unfinished*, The Urban Book Series,
https://doi.org/10.1007/978-3-030-73457-2_4

also involving our corporeity. Leftover reuse design could set at its centre the people who live in places, overcoming formal and purely visual principles on which interior projects have often been founded. This practice can assume a topical role as an answer to the urgent need of ceasing soil consumption, promoting urban regeneration and social innovation.

In the paper, different aspects of contemporary scenario will be dealt with, starting from an ecological perspective, exploring empathy and synaesthesia, focusing on the concept of atmosphere and concluding with a spotlight on the reuse of historical spaces.

4.2 Obsolescence and Soil Consumption

According to Dal Borgo et al. (2016), the landscape that best expresses contemporaneity is that of abandonment, which seems to be related to an interruption of perspective and possibility, a suspension, a void of cultural sense.

After the industrial revolution, the earth started to be exploited and worn out, disregarding the consumption of non-renewable energy sources. This led a small portion of humanity to great prosperity and wealth, sustained by contemporary great poverty in the rest of the world and an even worse catastrophe from an environmental point of view. The other side of the neo-liberalistic coin has been a symbolic impoverishment, a loss of environmental and cultural properties of places, and the progressive deletion of geo-aesthetic and anthropological *genius loci* (Marini 2015).

Before the advent of mass industrialization, things had to last and therefore were repeatedly repaired. In contrast, industrialization introduced the concept of obsolescence that substituted repair with replacement. If places are the mirror of the society's soul, a constant process of urbanization and cemented environments are the expressions of the apparently unlimited growth of Western countries. Standardization affected also places that are being used and thrown away when, for whatever reason, they become obsolescent.

According to Latouche (2012), "the obsession to sell" approach of economic culture and the advent of disposable good conceived the idea of planned obsolescence. Thus, to continuously sustain the capital cycle, the products are created to be consumed in a short time and then bought back.

This process affects places and landscapes too. The planned obsolescence of places creates rubbish, transforms territories into landfills and increases the wastage of natural resources exponentially. Today, the impact of human actions has reached a level that interferes with the great cycles of the planet's ecosystem.

The ground pollution and sterilization due to the accumulation of obsolescent places and waste production prevent the regeneration of soil as a resource. According to Pileri and Biondillo (2015), every hour in Europe, about 11 hectares of land are consumed by urbanization processes. This exponentially growing cementification has resulted in soil sealing, compromising its functionality and potential. By reducing the surfaces dedicated to food production in favour of surfaces devoted to mass production, those used for the former function are overexploited, causing them to decay due to excessive effort.

The soil is a very thin fabric and, like any epidermal tissue, it demonstrates the health of the surrounded organism. Like the planet's skin, it protects from atmospheric variations and temperature changes, regulates all relations between interior, surface and exterior, and is a laboratory of energy and raw materials that gives life to all that is above it (Pileri and Biondillo 2015).

One of the main problems of our era is reflecting on the problem of abandoned places and interrupted landscapes, understanding that, above all its possible endogenous and exogenous causes, there is a cultural posture which might allow overcoming both natural and artificial causes (Figs. 4.1 and 4.2).

Fig. 4.1 Dismissed factory, Hafen-Drehbrücke, Krefeld, 2019

Fig. 4.2 Dismissed psychiatric hospital, Limbiate Milan, 2017

4.3 Space Fascination

Human beings live embedded in a landscape that they perceive through their whole body. Their ecological relation to the environment affects their well-being and is at least as essential as the brain to the existence and exercise of the mind. In fact, knowledge begins with perception, and perception is perceiving the environment (Mace 2005).

The dependence of cognition upon the body and its environmental interactions is explained by saying that cognition, in real biological systems, requires some kind of information processing which involves exploration and modification of the environment. Indeed, the gaining of knowledge is a way to guide behaviour in response to the system's changing surroundings (Milkowski 2013).

According to Menatti and Casado da Rocha (2016), health is greatly affected by a landscape which represents something more than just part of our heritage or a place to be preserved for the aesthetic pleasure it provides. Rather, we can talk about the right to landscape as something intrinsically linked to the well-being of present and future generations. To explain the health-landscape relationship, they suggest the introduction of the concepts of agency and *affordance* in a new theory of landscape. In order to understand what landscape is and how it affects our health, we have to address biological and ecological theories of perception of the environment.

In their theory of affordance, human beings play an active part in the process of co-creating landscape, both from a cultural and an ecological point of view. Life, body and perception cannot be detached from the landscape on which relies people's mental and physical health. This agential dimension of the notion of the processual landscape seems to provide the connection with health and well-being. Indeed, the intimate connection between our body, our health and the landscape in which we live, or at which we look, has been demonstrated from different points of view (Menatti and Casado da Rocha 2016).

The term *topophilia* has been introduced in the literature and then mentioned by Bachelard in his *La Poétique de l'Espace* (1958) to indicate a cultural, biologically based, close connection to place. Topophilia (from the ancient Greek *topos*: place and *philia*: love, attachment) means that human feelings, values and attitudes towards the world are geographically "embedded", implying that experiencing places plays a fundamental role in our development. In modernity, such connections have been mostly destroyed but recently they are trying to be more and more reinvented.

Important contributions to this debate came from the field of human geography, when Tuan (1974) provided an understanding of place as a product of perceptive and cultural elements: place and perceiver are linked by values, ethical commitments and feelings. He also introduced the idea that anonymous space is changed into articulated geography through the actions and values of people. His distinction between space and place and his genealogy of the concept of place have become very important for many geographers, philosophers and sociologists, as well as for cultural approaches to place in general. His idea of topophilia has also been studied in connection with well-being, meaning that individual preferences for specific places and restorative environments are significantly associated with the quality of life (Menatti and Casado da Rocha 2016).

Interesting investigations have been carried out on the qualities that characterize restorative environments, i.e. environments that help restore our attention. Attention fatigue has been recognized as resulting in irritability, anxiety, stress, lack of perception and lack of interest in human beings. When our directed attention fades, our focus needs to be maintained through alternative modes of attending that would make the use of directed attention temporarily redundant. The solution lies in *fascination*, which has been described as "attention that requires no attention at all, such as when something exciting or interesting happens and we look to discover what is going on" (Menatti and Casado da Rocha 2016).

A theory has been developed on the basis that we can better concentrate and restore our directed attention after experiencing nature, which is a non-threatening environment that reduces the physical indicators connected with stress. Natural environments turn out not only to reduce stress by eliminating directed attention fatigue, but also to help preventing it. Based on this psychological theory focused on the restoration of attention and the involuntary role of fascination in the human mind, we can make the hypothesis that fascination comes not only from natural landscapes but also from historical, existing, decayed manufacts, rich in stratified memory.

4.4 Places and Corporeity

The neuronal mechanisms of our body explore and evaluate the emotional value and accessibility of our surrounding environment through the embodied simulation of materials, models, spatial relationships, sounds, smells, tactile qualities, scales, textures, patterns and atmosphere. Architecture generates multisensory impressions, so it seems to involve "mirror neurons" more than other arts.

According to Mallgrave (2013), to understand our experience of architecture, the role played by emotions is fundamental. Subjectivity consists in seeing the world from a perspective situated in time and space, characterized by the motory potential of our body and a more or less satisfying emotional colour. Looking at a building, a room or a design object also means emulating the movements and actions that those spaces and objects evoke. Mallgrave criticizes contemporary architectural design and considers it incapable of proposing buildings on a human scale by virtue of its exaggerated predominance of formal and pure visibilist criteria, with the consequent suppression of the other senses, such as touch, and more generally of the users' corporeality. The body does not only have a physical dimension: it is enriched with dream and memory, past and present. Indeed, memory is built not only through our nervous system and brain, but also through our bodies: the senses constitute a fundamental element of mediation for the information and judgement of the mind.

The fact that we perceive and therefore conceive the built environment through our whole body, and not simply our senses or our brain, is not obvious; architects tend to think of buildings as abstract objects or formal compositions that exist in a free geometric space rather than as existential places of our tactile consciousness. With the bodily knowledge we have of the world, sometimes metaphor seems the only possible way to state something about our non-physical experience: emotions, affections, feelings and moods. The body is so essential to thought that we can say that the mind is incorporated (Mallgrave 2013).

4.4.1 Empathy and Synaesthesia

Perceiving, choosing and using an object is the conclusion of an empathic relationship; empathy is the English translation of the German Einfühlung, a term introduced by Theodor Lipps to describe the relationship between a work of art and its observer.

Similarly, affordance is that set of actions that an object invites to perform on itself. Affordances invite us to use the objects, so in understanding those invitations we enter into empathy with the objects, which allows us to live and share emotions, experiences, needs and objectives (Crisci 2012).

Through their aesthetic aspect, objects and spaces play an emotional role, capable of evoking positive sensations. Managing shape, material/grain/texture, colour, value, transparency, size, position and orientation may result in a considerable activation of mirror neurons. In fact, the analysis and understanding of perceptive variables have an active part in the design process, as elements responsible for empathy.

Thanks to neuroscience, it is well known that some areas are activated in a similar way in all human beings when they experience identical emotions. Among the many disciplines that are formed on the different contributions of neuroscience, neuroaesthetics comes close to design in understanding what triggers and makes us judge the world aesthetically pleasing. The main function of mirror neurons is not only imitation, but also the understanding of the purpose of the action, in practice the intentions of the other. Mirror neurons do not reproduce the sensory perception of pain or discomfort, but a visceral sensation, which is experienced at the first approach with an artefact. Through mirror neurons, we experience in our body the emotion of the others, which becomes our own emotion.

Mirror neurons seem to explain the scientific evidence of empathy, already prefigured by the exponents of phenomenology. Wittgenstein wrote: "We see emotion (…) We do not see facial contortions and make inferences from them (…) to joy, grief, boredom. We describe a face immediately as sad, radiant, bored, even when we are unable to give other descriptions of the features" (Crisci 2012).

Perception, action and understanding cannot be considered independent of each other, just as the motor act is not considered only as an executive but also as a cognitive capacity. The theme of motor reactions can easily be traced back to the Heideggerian theme of usability and the Gibson theme of affordances, i.e. relationships and invitations provided by static or moving objects, which convey messages about their possible uses, actions and functions: "each object reveals its essence (…) a fruit says *eat me*, water says *drink me*, thunder says *fear me*, the woman says *love me*" (Crisci 2012).

In order for an object to be fully perceived, it must allow the body to acquire information through all senses: sight, hearing, smell, taste and touch; for this reason, the aesthetic experience involves people synaesthetically. In fact, the holistic nature of perception tends to grasp the sense of things by evaluating them on several levels. As described by Merleau-Ponty, every sensation perceived by an organ undergoes a synchronic transformation and processing through the complex of senses that we define synesthesia (Crisci 2012).

4.4.2 Atmosphere

The success of a place, from a room to an open space, is dictated by its ability to create urban atmospheres and seductions (Crisci 2012).

According to Norberg Schultz (1980), the place is an integral part of existence, a space with a distinctive character. It is made of concrete things with their material substance, form, texture and colour. All together these things define an "environmental character" which is the essence of the place, or its "atmosphere" (Norberg Schultz 1980).

In the context of philosophy, geography and architecture, atmosphere research is a growing area addressed to the emotions we feel in natural and artificial environments across day to day life (Trigg 2016).

Atmospheric feelings are felt with the body: worries strangle the throat, sadness weighs down the heart, an object of admiration almost cancels out the distance that separates us from it, an object of a reverential fear pushes us to withdraw and leave it a suitable space of manifestation and the vaults of a gothic cathedral invite us to lower our eyes and make silence (Griffero 2014). The atmospheric coldness of a room, which can be traced back to the type of furnishing or sound, to the colour of surfaces, or to the temperature, the light, the voice's timbre of the persons in it implies synesthetic involvement. A sound is velvety, strident, voluminous, earthy, cavernous, amber-coloured, polished in the light of an atmospheric synaesthesiology so unarbitrary that it can be artificially generated (Griffero 2014).

Atmosphere is something more than a sum of elements; it infuses pressure that can be positive and/or negative, of fear or joy. Thanks to our senses, we can locate the atmosphere in single aspects. The atmospheric dimension can be defined as the domain where the experiential vocation of architecture takes place. Based on the physical nature of the built environment, it then goes beyond it. In fact, atmosphere concerns both the measurable field of the physical parameters of the built environment and the ephemeral one of personal feelings. Overall, atmosphere corresponds to the impression marked on our senses and intellect by the experience of architectural space (Figs. 4.3 and 4.4). Though the atmospheric aura is not instantly perceivable, architecture cannot be separated from it. As a state, atmosphere is hardly defined not because it is rare and unusual but, on the contrary, because it is omnipresent, whether people are fully conscious of this or not. "Atmosphere is a complex phenomenon because it is invisible, intangible, elusive, without physical limits, unstable, instinctive, highly subjective, and often described through metaphors" (Canepa et al. 2019). Given the difficulty to deal with the atmospheric issue and to find a clear understanding due to its impalpable nature, many studies tried to explore it from an architectural perspective, by studying the topic through a sensory-emotional filter of the perceiving subject located in the built environment. According to Pallasmaa, atmosphere results from an exchange and fusion between the space and the subject, between the material or existent properties of the place and the immaterial realm of human perception and imagination. Yet, it is not a matter of physical things or facts, but of human experiential creations. Paradoxically, the atmosphere is grasped before its details are identified

Fig. 4.3 Atmospheric quality in the cathedral of Lucca, 2020

or it is intellectually understood. In a sense, it is a diffusely felt multi-sensory image, a singular experience that is blended with our very existential experience and cognition (Pallasmaa 2014).

4.5 Reusing Leftovers

Post-industrial cities are marked by the presence of underutilized complexes, buildings, urban landscapes and voids, contemporary ruins, areas affected by natural disasters or armed conflicts that form a complex palimpsest of memories, a system of relationships to be reinterpreted and redeveloped (Anzani et al. 2019).

Often, abandoned urban spaces constitute urban voids "perceived as an elsewhere", experienced by the population as an absence of meaning, a sort of missed opportunity for the city. Sometimes, historical emergencies, occasionally characterized by valuable constructive and compositional features even if in a state of abandonment, are part of a recognizable urban imaginary and a collective memory to which the population feels very attached.

Fig. 4.4 Atmospheric quality in the historical centre of Gdańsk, 2018

The ecological issue related to the need of ceasing soil consumption, many suggestions coming from different platforms and researches on urban regeneration and social innovation indicate the reuse of what the industrial city has abandoned as an extremely topical practice. The reuse currently concerns assets in continuous and progressive dismantling, linked to the socio-economic, functional and spiritual changes of our society, like industrial archaeology, military districts, former psychiatric hospitals, churches and religious complexes. According to Caramel, the term reuse has always raised questions at the centre of the architectural debate, especially if referred to the re-functionalization of pre-existing resources that were particularly significant for the community. The concept opens up to new possible interpretations that must necessarily be related to the complex transformations taking place in today's society. Giving back a sense to the different materials of the city, enhancing the potential meaning and identity of places, can take place at the level of a re-signification process of the space (Caramel 2020) through a multidisciplinary approach.

Adopting minimal intervention criteria, as well as hybridization and reversible approaches in the preservation of industrial icons, in the maintenance of the intrinsic historic character of ancient villages or dismissed public complexes, may achieve different goals, from the enhancement of local identities, to the redevelopment of heritage tourism, favouring individual and community comfort (Anzani and Caramel 2020).

Interior design addressing the reuse of abandoned buildings should recognize their meaning, their historical and material stratification and the added value deriving from the reciprocal enhancement between the new and the old. In many cases, the greatest benefit deriving from the reuse of an existing building is the possibility to appreciate its delicate and often hidden beauty revealed through the project. Beauty safeguard and enhancement are actually a real contribution to improving life quality.

4.5.1 Historical Places

The intimate quality of a place is due not only to the perception of climate and geography, but also to imagination and memory. "Comforting and inviting settings inspire our unconscious imagery, daydreams and fantasies" (Pallasmaa 2014).

Buildings and cities do not create simple objects of visual seduction; they get into relation, mediate through the senses and convey meanings. This hypothesis suggests that one of the reasons why contemporary spaces often alienate us—compared to historical and natural settings that arouse strong emotional involvement—has to do with the poverty of their impact on our peripheral vision and the consequent weakness of atmospheric quality. Peripheral perception is the perceptive mode through which we grasp the atmospheres, transforming retinal images into a spatial and bodily involvement and giving rise to personal participation.

Dwelling on the concept of fascination mentioned above, we can make the hypothesis that this could come not only from natural landscapes but also from historical, existing, decayed manufacts, rich in stratified memory (Figs. 4.5 and 4.6).

Ultimately, while the visceral aspect of the design is interpreted only and uniquely by the senses, the behavioural and reflective aspect requires the use of reasoning and is therefore linked to experience and memory.

Research in environmental aesthetics shows that people generally prefer historical places to modern architecture since they create a sense of continuity with the past, embody the group traditions and facilitate place attachment, intensified by the awareness of the place history (Lewicka 2008).

In fact, citizens need a sense of belonging. According to a cultural principle of identification, we can speak of cultural territory by defining it as a form of rootedness and attachment to places. Communities do not build their environment for a material purpose, but to offer an image of themselves, to become aware of what is a common good and to experience the relationship with others (Amoruso and Battista 2020).

4.6 Conclusions

Reuse design of leftovers can be seen as a practice which promotes individual, community and planet well-being. It should set at its centre the people who live in places, overcoming formal and purely visual principles on which interior projects

Fig. 4.5 Reuse project of a historical complex in Milan (Interior Design Studio, Politecnico di Milano, 2nd year project by Boldrin L, Cartagena D, Cioni I, Cipriani V, Dalmolin D, De La Vega Rueda G, Terruzzi C, 2020)

Fig. 4.6 Reuse project of a historical complex in Milan (Interior Design Studio, Politecnico di Milano, 2nd year project by Allioli I, Cangelosi C, Mazzina B, Parenti E, Rossi F, Somaschini A, Sozzi E, 2020)

have often been founded. Indeed, in addition to problems due to urban density and soil consumption, the recent Covid-19 pandemic has proved megacities to be particularly inadequate to prevent contagion diffusion and to provide healthy psychophysical conditions. Different indications have been given, both at a private and a public level, to address reuse experiences of small abandoned historic centres as well as of abandoned urban assets so as to favour neighbourhood relations and proximity experiences. In this perspective, the design culture can play a key role, by adopting innovative and original design thinking, welcoming a transdisciplinary perspective which could take advantage of the recent application of neuroscientific discoveries to the architectural and interior design field. The challenge is to experiment interventions able to return dismissed areas to new use possibilities (Crespi et al. 2017), relying on temporary and reversible devices, consistent with the place nature and soul, aimed at their reintegration into the living social fabric and the enhancement of their symbolic significance and capable of involving people in synaesthetic experiences and engaging atmospheres.

References

Amoruso G, Battista V (2020) Landscape. Bottom-up approach for cultural landscape and local identity mapping. In: Anzani A. (ed) Mind and places. A multidisciplinary approach to the design of contemporary city. Springer, Cham, 223–236

Anzani A (2020) Identity. Place identity between preservation and innovation. In: Anzani A (ed) Mind and places. A multidisciplinary approach to the design of contemporary city. Springer, Cham, pp 267–274

Anzani A, Caramel C (2020) Design and restoration—an ecological approach. In: Crespi L (ed) Cultural, theoretical, and innovative approaches to contemporary interior design. Hershey, Pennsylvania, IGI Global, pp 68–84

Anzani A, Caramel C, Lonardo E (2019) Hybridization and reuse of existing buildings. In: Scullica F, Elgani E (eds) Living, working and travelling: new processes of hybridization for the spaces of hospitality and work. Franco Angeli, Milan, Design International series, 139–148

Bachelard G (1958) La Poétique de l'Espace. Presses Universitaires de France, Paris

Canepa E, Scelsi V, Fassio A, Avanzino L, Lagravinese G, Chiorri C (2019) Atmospheres: feeling architecture by emotions. preliminary neuroscientific insights on atmospheric perception in architecture. Ambiances, 5

Caramel C (2020) Reuse. Urban heritage and liveability. In: Anzani A (ed) Mind and places. A multidisciplinary approach to the design of contemporary city. Springer, Cham, 245–254

Crespi L, Anzani A, Caramel C, Crippa D, Di Prete B, Lonardo E (2017) Designing remains. In: Amoruso G (ed) Putting tradition into practice: heritage, place and design. Cham, Springer, pp 1473–1482

Crisci G (2012) Specchi del design. Una ricerca sull'empatia fra teorie estetiche e neuroscienze. Master thesis, Design Department, Politecnico di Milano, tutor S. Zingale

Dal Borgo A, Garda E, Marini A (2016) Sguardi tra i residui. I luoghi dell'abbandono tra rovine, utopie ed eterotopie. Mimesis, Milan

Griffero T (2014) Atmospheres. Aesthetics of emotional spaces. Ashgate Publishing, Farnham

Latouche S (2012) Bon pour la casse! Les déraisons de l'obsolescence programmée. Les Liens Qui Liberent Editions, Paris

Lewicka M (2008) Place attachment, place identity, and place memory: Restoring the forgotten city past. J Environ Psychol 28:209–231

Mace WM (2005) James J. Gibson's ecological approach: perceiving what exists. Ethics Environ 10(2):195–216
Mallgrave HF (2013) Architecture and embodiment: the implications of the new sciences and humanities for design. Routledge, Abingdon-on-Thames
Marini A (2015) Paesaggi interrotti. Geostoria e geofilosofia dei luoghi abbandonati. PhD thesis, Università degli Studi di Milano, PhD course Scienze dei Beni Culturali e Ambientali XXVIII Cycle, tutor L. Bonardi
Menatti L, Casado da Rocha A (2016) Landscape and health: connecting psychology, aesthetics, and philosophy through the concept of affordance. Front Psychol 7:571
Milkowski M (2013) Explaining the computational mind. MIT Press, Cambridge, MA
Norberg Schulz N (1980) Genius Loci. Towards a phenomenology of architecture. Rizzoli, New York
Pallasmaa J (2014) Space, place and atmosphere. Emotion and peripheral perception in architectural experience. Lebenswelt Aesthet Philos Exper 4
Pileri P, Biondillo G (2015) Che cosa c'è sotto. Il suolo, i suoi segreti, le ragioni per difenderlo. Altraeconomia, Milan
Trigg D (2016) Atmospheres, inside and out. Environ Plan D Soc Space 34(4):763–773
Tuan YF (1974) Topophilia: a study of environmental perception, attitudes, and values. Prentice-Hall, Englewood Cliffs, NJ

Chapter 5
Leftovers at the Start. From the Analysis of the Theme to the Development of Design Strategies

Emilio Lonardo

Abstract To demonstrate a theory, to sediment contents or to spread a manifesto, there is no more fertile ground than the University. Students are at the same time the lifeblood that transmits energy to new things, and the litmus paper to understand if the elements proposed have the characteristic of being feasible in the contemporary context. In the last five years, the 'unfinished' as a didactic topic has been proposed in the 'Laboratorio di Sintesi Finale' of the School of Design of the Politecnico di Milano, entitled Avanzi. Working on territories and buildings of a different nature, linked by the condition of Avanzi, in other words, spaces that have been built to perform a function and that are then advanced, as when you cook in excess. The Laboratory provides that the students put into play all the notions learned during the three years of the course of study, deepening the project, making it their thesis degree. Among the objectives (and methods) of the didactic approach, it is required to develop design strategies that can be applied to the theme. The chapter is divided into two main parts, the first one is a reconnaissance on the topic, the second one is a luggage of strategies presented to the students as an example and then some of the ones provided by them are presented.

Keywords Studio · Strategies · Didactic · Unfinished · Leftovers

5.1 Introduction

> What's not ridiculous is what's in between, the time you wait between sex and love facing the leftovers of what's left with a sense of goodness … without becoming bitter. I think that we must become good with the little is left within us, what is left over after it has been discarded. In other words, staying in one piece even when things didn't work out. I think you need a little luck, a little glamor, a little strength and a little coolness to continue living. […] Cold blood means moving forward when everything seems terrible. […] It means getting by when everything seems so terrible that it's not worth continuing, and you don't turn to God, you don't go to church. Face the wall you do it yourself. If you think it's not hard, honey …

E. Lonardo (✉)
Politecnico di Milano, Dipartimento di Design, Milan, Italy
e-mail: emilio.lonardo@polimi.it

L. Crespi (ed.), *Design of the Unfinished*, The Urban Book Series,
https://doi.org/10.1007/978-3-030-73457-2_5

> this is hard, honey. Running somewhere, holding on to something, a god, a woman, a drug, a successful evening, for a night, a week, a year, a life; people don't sit still long enough to figure out who the hell they are.

This is how Charles Bukowski told in an interview entitled 'Statements of a dirty old man' with Silvia Bizio in 1981 and published in the collection 'The sun kisses the beautiful: interviews, meetings, insults' about his relationship, sometimes dramatic, with the things of life.

And just like people, often, we also meet places that need to be saved; they can be personal, intimate places or pieces of the city, they can make us feel safe, or oppress us, but surely, we feel them ours, our places, anthropic and influenced by our choices, as in turn, they influence our emotions and our behaviors.

Guy Debord, director and writer had already guessed that, by referring to his research and experiments *he* carried out within the Situationist International and culminating in the publication of Naked City, that is a cartographic reconstruction of Paris *he* obtained from the recomposition of nineteen sections of the city. *He* summarized with the term Psychogeography and defining it as an area that has the task of investigating the relationship between people and the built, the study of specific laws and effects that, consciously organized or not, the geographical environment has on emotions and behaviors of individuals (Debord 1955).

Human beings are crossers of spaces, who continually mark with the traces of themself over the months, years, centuries, but also at the end of every single day, every piece of the world they traveled, which is enriched with new layers of stratification. They are the layers of the memory of places that often tell the inner path of a single person's life, as in 'Merzbau' by Kurt Switters and they resist, though naked, poetic and diaphanous, to the signs of time, to the scratch of the most powerful catastrophes, as in Mona Hatoum's work 'Remains of the day'.

And in this world, made up of spaces and places, intersections of experiences, humans move to achieve a goal or a place, by using the shortest route, cutting out oceans of possibilities and obeying social needs with an anthropological foundation.

As complementary opposites, these needs include security and openness, certainty and adventure, work organization and fun, predictability and the unexpected, unity and difference, isolation and encounter, exchange and investment, independence [or loneliness] and communication, immediacy and long-term perspective. The human being also has a need […] for creative activity, for works [not only for consumable products and material goods], a need for information, symbols, imagination, playful activities. A fundamental desire lives and survives through the specific needs, of which play, sexuality, bodily activities such as sport, art and knowledge represent particular manifestations and moments that allow to overcome the specialized work division at different levels (Lefebvre 1976). Work is at the center of Ettore Sottsass Jr.'s reflections who denounced in 1973, by imagining the 'Planet as a Festival', those 'uninvolved thoughts on cities' according to which human beings must live to work and work to produce and then consume; the architect and designer proposed to react so that they can instead live to live and work, if they want, 'to know with their body, psyche and sex they are experiencing'. In Sottsass's considerations, we can see a clear reading of the changes/transformations that were affecting the society of that time, more and more global.

As capitalism and the urbanization process take root, the value of exchange as an 'object' replaces its use value, depriving citizens of the fundamental meaning of urban life and the city as a place of participation, encounter and interaction, actually changing the very meaning of the urban place. The setup of these factors due to the economic crisis and technological development has provided a complete and immediate access to a huge number of resources, even in mobility. Furthermore, the decrease of time and costs for medium and long-range travel is redesigning people's life who tend to nomadism and so reconfiguring the cities as global networks nodes. However, almost every analysis seems to confirm that the fragmentation, dispersion and heterogeneity of the contemporary city can be attributed to the technical process in the field of transport and telecommunications according to a model that aims to explain the urban development with regard to Kondratiev's waves.[1]

The contemporary city, rather than a place where one lives and inhabits, has become a non-place where one moves; it is a space to be crossed, a territory where subjects keep moving by creating magnetic fields that define it beyond the material aspects of which it is composed. This is the antithesis of the definition of traditional city, which was a strong, fixed, material point. The city was a place to go, a destination where to operate. The opposite of a territory where to move (Poli 2009).

If you consider that the changed relationship with happiness is revealed by the generational leap between baby boomers and millennials, it's possible to draw a fairly clear line with respect to the direction taken on the use of cities. The socio-demographic surveys indicate—in fact—that the children of the economic boom feel happy when they can satisfy material goods such as having a nice house, a good job, a nice car, in the contrary the ones born after 1980 get the maximum satisfaction from traveling, knowing, exchanging. In the space of three generations [in-between there is the one called X], a revolution has taken place: possession has stopped being the main road to personal fulfillment to leave space for experience (Botti 2017). This new type of nomad does not know how to say what is 'home' and what is 'family'; they replace relationships, groups, and places to live in a sort of exponential consumption. We are talking about an individual which is formed as a trans-cultural hybrid, equipped with a fair economic strength that has generally acquired [and not gained], the ability to easily cross multimedia, cultural and political borders; a 'homo visitor', a collector of shots and experiences, rather than a subject who has an interest in bringing added value to the public sphere. And if this is the evolutionary stage that the humans are reaching, then one wonders/we wonder what cities will be used for, and who will take care of them, if there will be social groups capable of reacting to the gigantism of global uniformity.

Actually, the contemporary world is the result of a strong simplification and an increasing inflexibility. The described scenario also leads to the impoverishment of

[1]Sinusoids capable of describing regular economic cycles within the modern capitalist system. Since 1771, five waves have been recognized: the Industrial Revolution, the Steam and Railway Era, the Steel and Heavy Engineering Age, the Oil and Automobile Era and the Telematic Age.

flora and fauna as well as languages and cultures, and also techniques and architectures. In the last 30 years, we have been moving from modernization to modernism and, right after the postmodern crisis, we have entered the vortex of globalization that has brought us, in the space of just a few decades, to a situation that is already post-global. But since 'if people as individuals do not go forward or do not strive to go forward, they go backwards and still and quietly cannot remain without corrupting themselves' (Croce 1992), Benedetto Croce says, today we must look for the new direction of progress. The claim of the right to the city will then have to pass through the reconquest of two key principles: one's own identity as a citizen and the consequent possibility to inhabit the city; and if in the Situationist motto, 'To inhabit is to be everywhere at home', later taken up by Ugo La Pietra as a slogan for much of his artistic and design research, we can glimpse the connection and the fundamental role that the domesticity theme can play in the rehabilitation of the urban context, then we argue that the innovations in the coming decades are from particular and deliberate communities. Asserting the right to a delimited geographic identity would represent a radical change after centuries of a homogenization trend, culminating in globalization. The scene for these groups is and must be the city (Poli 2009).

As in the movie 'Synecdoche, New York', the protagonist, a theater director, stages a show that turns into his own life, then the city becomes a projection of society onto the territory. Today's society has generated new social categories and inhabitants use spaces in unusual ways. So, if on the one hand the history of design still offers timeless footholds interested in the objects which are designed to last, on the other hand you would like to find a proposal in order to meet new inhabitants' needs of contemporary cities. You could think of works such as the collection of suitcases designed by Marc Sadler for the company FPM that are containers able not only of carrying clothes, but also of turning into mobile offices or even kitchens. But if you look at projects like Munari's Abitacolo, you can't help but be fascinated by the poetic mastery, not so much of the sign, but of that subliminal attitude that makes us perceive its quality. Abitacolo is not only an object, but also a space to express a clear light materiality that builds its aesthetics; at the same time, it is adapted to the needs of the user, who is not submitted to it, but *he* lives it, by creating unique and different experiences.

5.2 The Dynamics of Experiences

The experiential component assumed a new importance in the context of contemporary design, precisely because of the paradigm shift in the way people see possession where the sharing economy is only one of the manifestations. Design has been influenced by this change. Today we are talking about a complex subject that does not end with the design and creation of consumer goods, but includes the immaterial sphere, working on the meaning of the contact experience with objects, companies, organizations or people. For a designer, it means changing perspective and putting oneself in the shoes of the user before those of the customer. It also means considering one's own limits and accepting the interdisciplinary dimension of an action that must combine tools and concepts from very different sectors (Botti 2017). Being able to design experiences means adding a sophistication level to the design of a space, and to do that, you need to know the main characteristics of the dynamics that are activated by living any kind of experience. First of all, experiences are personal, they take place inside the individual who is involved on an emotional, physical, intellectual, or even spiritual level. What's the result? Two people cannot have the same experience—at all. Each experience is a result of the interaction between the staged event and the individual's previous mental and living conditions (Pine and Gilmore 2000). We can identify four main clusters of experiences as follows: aesthetic, entertainment, educational, and escapism experiences.

5.2.1 Aesthetic Experiences

In an aesthetic experience, the individual plunges into an event while remaining passive. An example of this type of experience can be considered the installation entitled 'The Weather Project' and set up by Olafur Eliasson for the first time in 2003. The work, created for the Turbine Hall of the Tate Modern in London, aims to bring sunshine to a grey city like London. Visitors entering the Hall found themselves suddenly plunged into a sunset consisting of a semicircular screen that radiated the lighting frequency of the sun. To make the experience more fascinating, Eliasson had the idea of mirroring the entire ceiling, so that the circle which represented the sun would be completed and the height of the space was doubled. To prepare the exhibition, Eliasson also developed a questionnaire for Tate Modern employees that included questions such as, 'Has a weather phenomenon ever drastically changed the course of your life?' 'Do you think tolerance toward other individuals is proportional to time?' 'To what extent are you aware of the weather outside of your workplace?' In addition, there was a panel discussion on art communication, weather reports on abnormal weather events, weather statistics, and a series of essays on weather and space.

5.2.2 Entertainment Experiences

In entertainment experiences, the individual passively assimilates what happens through the senses. In this regard, the GAT Fog Party that was first designed by Martí Guixé for the CASCO space in Utrecht in 2004 can be considered explanatory. The project plays on the borderline between watching and consuming by enveloping visitors in a cloud of smoke generated by Gin and Tonic (hence the meaning of the acronym GAT) (Fig. 5.1).

Fig. 5.1 Gat Fog. Casco, Utrecht 2004. Design Martí Guixé. Photo copyright Inga Knölke

5.2.3 Educational Experiences

If an individual actively participates with his/her body or mind, absorbing the event, we will talk about educational experience. 'Dialogue in the Dark' is an activity that you can have at the Institute for the Blind in Milan, where the visitor has the opportunity to walk for an hour in total darkness and then experiences, at first hand, what means to live everyday situations, without the sense we consider the main and so, learning a new way of 'seeing'.

5.2.4 Escape Experiences

If, eventually, the individual is completely involved and actively participates in the event, then we are talking about escapism experiences. Let's think about stores where people have the opportunity to try out products before making a purchase; the Globetrotter chain, which specializes in the outdoors, has been carrying on this philosophy for years. For example, its headquarters in Munich that is opened in 2011 and designed to engage customers in a range of outdoor-related activities, such as trying out rain chambers, paddling in a canoe, or testing high-altitude equipment and clothing in the 125 m^2 cold chamber, in which weather conditions that would be experienced during a trek have been simulated.

Over the past decade, it became more and more clear that a space designer must also take into account the experiential component of the spaces to design. Thus, part of the design process also involves the placement of experience. If you imagine a four-quadrant diagram, then you can try to figure out what kind of experience to try and play out, if you should lean toward one or create a balanced experiential design.

In this scenario, even the material component can be questioned, to promote a dimension that is always part of our daily lives, but often little considered during the design phase: the time (Fig. 5.2).

Carlo Rovelli, in his book 'L'ordine del Tempo' (The Order of Time) writes that 'The world is made of networks of kisses, not of stones' (Rovelli, 2017); this concept perfectly connects the theme of experiences with a design matter as important as difficult to shape. Time in fact, is a physical, material component; for instance, let's think when we measure great distances among planets in light years, with time. For this reason, it is possible to say, in some way, that time is also a space. Without going into all the existing theories on time—an entire book would not be enough, and especially more qualified people would be needed to do it—anyway, we can quite simply say that usually we are used to divide time into three macro moments: the past, to which belong memory and recollection as well as remembrance; the present, which is the space of the context, what actually exists; and the future, which is the place where the project is placed. Moreover, our everyday life is filled with

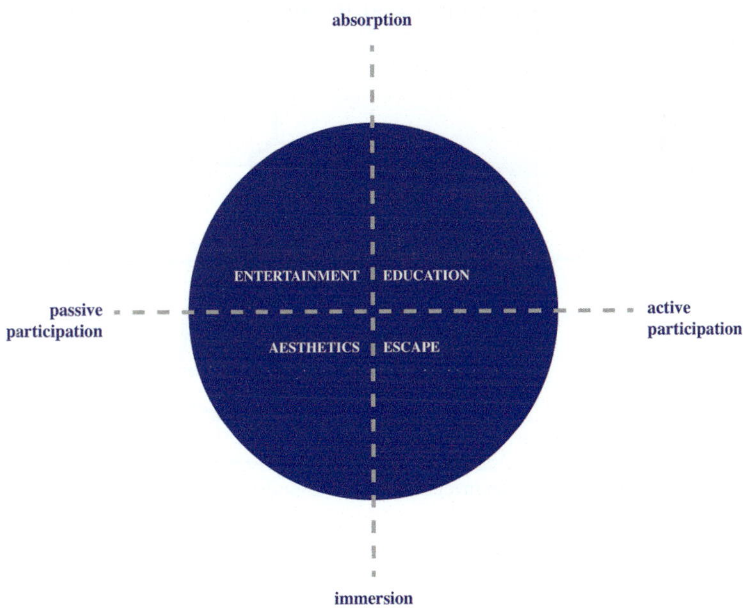

Fig. 5.2 Four-quadrant experiences diagram

a big misunderstanding; we measure time with accurate and repetitive mechanical instruments, compared to the complex nature of time. In fact, there is a psychological perception of time that is often far away from its mechanical measurement. Almost everyone, at least once in his or her life, has experienced a dilation or contraction of the time perception; just think of all the times we say, 'time has flown by' or 'this hour won't pass anymore'. It is in this gap that obsolete objects and spaces are created, it is in this crack that leftovers are placed.

5.3 Some Considerations on Leftovers

The designer of spaces has the opportunity, and maybe the duty of taking this gap to design new opportunities and give a new life to the forgotten places. At the end of the 1970s, Ugo La Pietra had already recognized some possible urban leftovers, and tried to imagine their functional recovery with the project 'Urban equipment for the community'; an example is represented by the telephone booths rethought as showers.

Before being able to talk about design strategies to act on the leftovers, it is necessary to consider some of their features and on how the designer's eye can see them.

The leftovers, in fact, represent first of all a distance which is created in the space of time and to be filled in in terms of architectural design. For this reason, also connected to what has been previously said, the first type of leftover to be considered is the temporal one (Fig. 5.3).

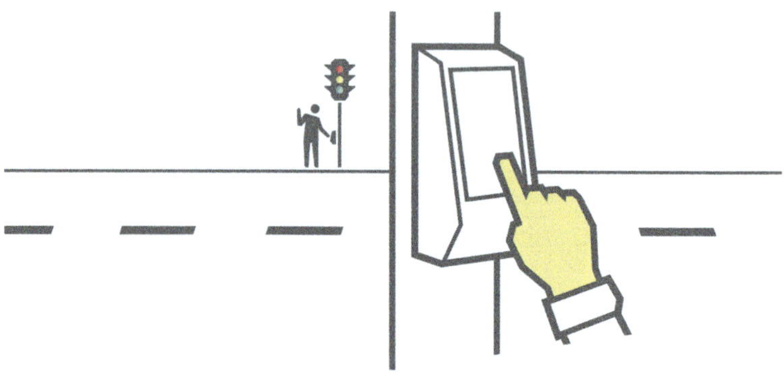

Fig. 5.3 ActiWait. Project by urban invention. Image Courtesy urban invention

5.3.1 Leftover as Time

In an ancient strip that appeared in Mickey Mouse and titled 'Mickey Mouse and the Tibetan device' on March 17, 1974, Gambadilegno and a scientist are talking about a machine to collect and store the lost time; the bus stop is the first place they think to find it. The time spent waiting for the bus is often considered as a lost time; I don't know if the artist Biancoshock read this comic strip thinking about his project 'Antistress for Free', but in fact, *he* fills that space in a creative way, that's placing a series of plastic sheets with bubbles of different sizes and according to the waiting time on some bus shelters in Milan; *he* uses a material that is usually seen as a waste, a surplus.

In the same way, it is possible to consider the time we spend at the traffic lights. Waiting is, in fact, usually experienced in a passive way, but, unlike waiting, it reveals an active feeling that stands for 'tending to'. In 2012, the Urban Invention duo designed the ActiWait system based on this principle, a device to be applied to city traffic lights in which it is possible to perform some active actions, such as playing pong with people who are waiting just like us for the green light to cross the street.

5.3.2 Leftovers as Meaning

Between 1915 and 1917 Marcel Duchamp created 'Fountain', by taking an object such as a urinal, removing it from its usual context, and generating in fact a leftover, which, however, took on a new meaning in its being ready-made; this translation of meaning was the genesis of all contemporary art. Even the leftovers we are dealing with are in fact, and in some way, from a functional or temporal point of view, out-of-context, and in any case they can be considered in a phase of standstill, balanced on a line that can declare their disappearance or rebirth. The designer who wants to work on the leftovers should have an approach that is not so different from the Duchampian one, able not to distort the physical nature of the space, instead managing to look at it and make it look, to give it, through the project, a new meaning.

5.3.3 Leftovers as Opportunities

An ancient Japanese saying goes: 'There is luck in what is left over'.

In the contemporary landscape that is full but also guided by a powerful drive for change, discovering places that are waiting for a second chance can be considered, in a certain way, a fortune for many of the actors who play a key role in decision-making processes. However, very often, administrations are short-sighted in front of

Fig. 5.4 Grindbakken, Ghent, 2012. Project by rotor. Image Courtesy rotor

the heritage they manage; in this regard, the skill of the designer is to recognize a surplus, understand its value and potential, while respecting its identity (Fig. 5.4).

The world—in fact—is scattered with unconscious and unnoticed 'works', that are signs with an aesthetic value, a meaning, but they are simply invisible; and some designers, thanks to their sensitivity, manage to highlight these parts of the world (Ciuffi 2012). When the area of the port city of Ghent was about to be reorganized by a new master plan in 2012, the Rotors glimpsed an unexpressed potential in a curious structure 160 meters long. Grindbakken, which had served as a deposit for gravel and sand, an intermediate step in the transfer from trucks to ships, after being dismantled, has been converted into an open-air exhibition space, and in which the white cube image has been interrupted and given rhythm thanks to the intervention of the Belgian collective. They decided to work on traces and signs on the surfaces, creating frames with negatives capable of isolating natural works, mixed with works by other authors, in a dialogue of languages that will gradually be lost under the action of time (Fig. 5.5).

On the other side of the world, in Japan, in Tokyo, the Atelier Bow-Wow, since 2002 coined the term 'Pet Architecture' to describe those types of buildings that are placed in the interstitial spaces of the urban context, producing a unique feature of ownership of the city spaces. The analysis of these architectures built on city leftovers, produced some of the most interesting projects of the Japanese group, such as the 'House tower' of 2006, which includes 9 juxtaposed rooms, connected by a central staircase in a volume of $3 \times 6 \times 11.5$ m.

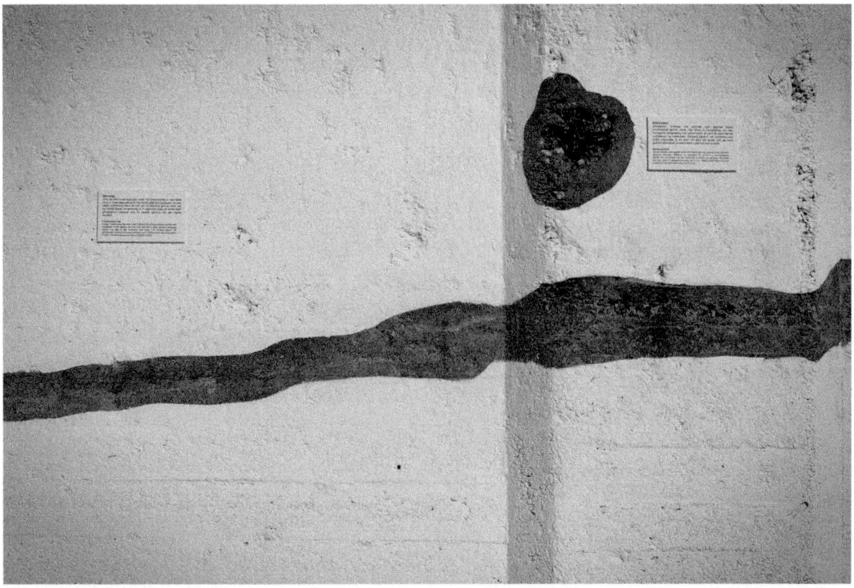

Fig. 5.5 Grindbakken, Ghent, 2012. Project by rotor. Image Courtesy eric Mairiaux

5.3.4 The Role of the Designer

In order to declare his intention to innovate, the designer should train his way of observing things, as well as listening to them. The skilled designer is the one who knows how to observe, listen and understand users that are also the spaces to (re)design, even before the clients. Ugo la Pietra knew it very well when *he* built the 'Commutatore', an object that can be considered as the manifesto of his research work on urban spaces (Fig. 5.6).

Many times, through its use, La Pietra was able to see things not immediately visible, many times *he* made other people use it. This tool is to know and also understand and propose, because reality is an intricate tangle that the trained eye of a talented designer can untangle and discover its full meaning, as happens to the viewers who find a privileged point of observation in the anamorphic works of Felice Varini. And once that sense has been found that same eye can see gaps to be filled, through creativity, a bit like the activities of Jan Vorman's Dispatchwork project, who travels around the world by filling the cracks in the historic centers with colored bricks. The role of the designer, especially today, is above all ethical, and his gaze should be inside, and not from above; the good designer should understand the responsibility of his own projects, undresses the dress of the archistars and wears what we could define as an 'anarchistar', who does not forget that his projects are not just empty sets for magazine shots, but places that are inhabited every day, influencing people's lives.

Fig. 5.6 Il Commutatore, 1970. Project by Ugo La Pietra. Image Courtesy Archivio Ugo La Pietra, Milano

5.3.5 A Luggage of Strategies

When the world of design, so intense, continues to impose itself more as a spectacular phenomenon, a pure skin arrogantly uninterested in quality (Crespi 2013), we should counter by offering different and timeless visions based on higher values. If on the one hand, therefore, companies guide their choices according to trends, the designer should have the responsibility to offer strategies, that is, lenses to see the world in a holistic way, by guiding transformation and innovation rather than following it.

The project, especially when applied to spaces, is a complex organism, made up of points of view and personalities that are sometimes very different. To borrow a metaphor proposed to me by the architect Attilio Stocchi, a project can be considered the meeting of several dreams. There is, for sure, the customer's dream, the one who commissions the work; then, the dream of the place, because places as well as people have a soul; there is the designer's dream who puts his vision into play; and finally the dream of the inhabitants, the ones who will live that space. A successful project cannot only partially take into consideration one or a few of these dreams, but also it must be able to keep them together, make them cohabit, mix them, make them fall in love with each other. In order to do this, it is necessary to develop strategies that are able to guide the design process (Fig. 5.7).

The following pages tell about, on one hand, some of the strategies the author developed to try and merge the research of emblematic case studies with his own experience as a designer, and, on the other hand, *he* offers an overview about some

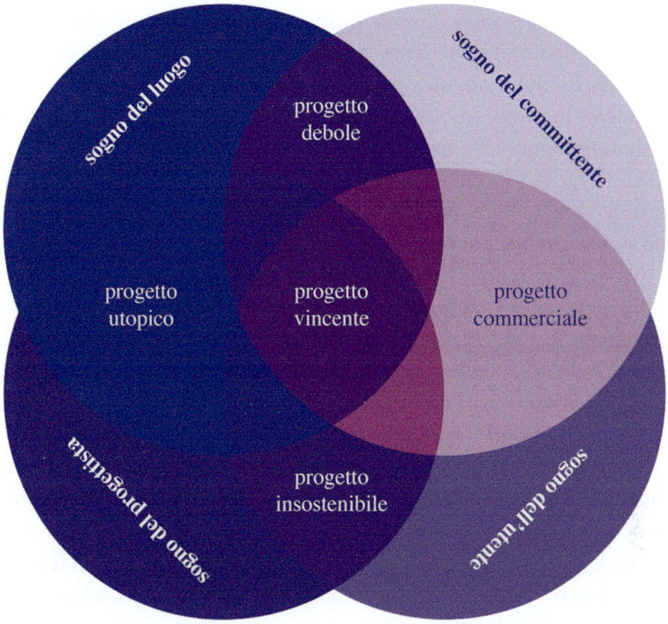

Fig. 5.7 The dreams of a project

of the most successful strategies developed by third-year students of Final Synthesis Workshop led by Luciano Crespi and also many of the authors in this volume, at the School of Design of Politecnico di Milano during the years when the theme of the workshop was Leftovers.

5.3.6 Long Time Strategy

In his book '25 ways to hammer a nail' Enzo Mari states that 'When people ask me who is the best designer I know, I answer: an old farmer who plants a chestnut wood. *He* doesn't plant it for himself, but for his grandchildren'. (Mari, 2011) This attitude tells of the virgin depth that every designer should be able to assume when approaching a new project, playing the role of farmer and grandfather at the same time, of professional and amateur.

In addition, this phrase also suggests to us the importance of the project's vision; designing for strategies means having a resilient vision, which can resist the fashions of time because it is able to adapt while staying true to its identity.

Every strategy, in order to achieve the objectives that are set from time to time, also needs tactics, implementation tools, and therefore it also presents a plurality of action shades.

Thinking about one's project over time can mean thinking about how it might appear after a certain number of years, as in the sensational representation elaborated by Joseph Gandy for John Soane's project of the Bank of England in 1813; but it can also simply mean thinking about how the project can adapt to the various seasons and establish a dialogue with time. Knowing the inner characteristics of a place, or a project you are going to create, allows you to think about how to extend its life; one of the most knowledgeable contemporary designers in this sense is certainly Paolo Ulian, who, descended to the scale of the object with works such as 'Double Match', a two-headed match that can therefore be used twice, or 'A Second Life' and a ceramic centerpiece with microholes that identify elliptical shapes capable of saving themselves in case of breaking the whole piece. In this way, he reaches a very high level of poetics and sophistication in thinking about leftovers as potential generative of new stimuli and new realities (Fig. 5.8).

To return to the scale of the space, and borrowing a photographic metaphor, one can also think of designing for long expositions, that is, offering spaces that are simultaneously unique and multiple, simultaneously impressing on the environment the appearance of its actuality and the nuance of all its possibilities, a bit like what happens in the design of public seating conceived by Ma0 for Piazza Risorgimento in Bari, an archipelago of possible configurations and relationships, adaptable according to the needs of the inhabitants.

Fig. 5.8 Una Seconda Vita, Ceramic centrepiece. Model realized for the Biennial of Ceramics Albissola, 2006. Project by Paolo Ulian. Image Courtesy Paolo Ulian

5.3.7 And Yet It Moves

As in the previous case, adding flexibility and movement components help to make a space somewhat alive and more able to establish a confident relationship with its users (Fig. 5.9).

There are several ways to apply this strategy to a project. One of them involves a physical/mechanical action by the people who want to modify the space, as in the case of 'All I own house' designed by PKMN architects for the graphic designer Yolanda R. Pila and in which, even in a small space, thanks to the combination of sliding systems applied to wooden and iron structures, all the domestic scenes of daily life coexist in response to the times and ways of its inhabitant.

A second possibility is to use motorized systems, as in the case of Renzo Piano Building Workshop's project for the Fondazione Emilio Vedova at the former Magazzini del Sale in Venice. The works are moved slowly along the length of the Magazzino and positioned at various heights in the exhibition space. Vedova's work is thus gradually revealed to visitors who can move freely among the canvases. Flexible and dynamic, the space is designed to adapt to different types of art exhibition. A third approach is to leave to soft components the burden of the movement action: wind, temperature, light are dimensions that, according to the conditions of the context, can generate vibrant effects in the space; let's think about Bruno Munari's 'Useless machines', or the series of art installations 'Plastic Bags' by Pascale Marthine Tayou

Fig. 5.9 All I own house. Project by PKMN Arkitects. Image Courtesy Enorme Studio

who use second-hand plastic bags, that's leftovers, objects 'in permanent transit, moving towards other destinations' and creates works that are caressed and moved by environmental factors.

5.3.8 I Need to Find a Sense

As seen above, one of the fundamental objectives of contemporary space design, especially in a reconversion of a Leftover, is to offer an experience as engaging and memorable as possible to the user. One of the ways to do that is to shape the sensory material in order to use it to generate unexpected sensations. While architectural practice has traditionally been dominated by the view, in recent decades an increasing number of architects and designers have started to consider the role played by the other senses, such as sound, touch (including proprioception, kinesthesia and vestibular sense), smell and, in a few moments, even taste. It is therefore clearly important to move beyond the purely visual focus (not to mention modularity) of architecture that has been identified in the writings of Juhani Pallasmaa and others (Spence 2020). In his book 'The Eyes of the Skin. Architecture and the Senses', Pallasmaa suggests that only an architecture that provides for a multi-sensory experience can be meaningful: a space that can be measured with the eyes, movement, muscles, and touch, that is, it realizes a co-presence of sensations relating the total perception of our bodies to the built environment (Pallasmaa, 1996). Then, through the activation of synaesthesia, the designer tries to design with the senses and for the senses, also creating different types of behavior. Ugo La Pietra, with the project 'Occultamento' (1974), creates a space in which the furniture system is hidden from view on the floor and in the walls, thus generating a space that is simultaneously empty and full, but the use of triggers that involve the other senses makes it to act at a subliminal level, allowing to design or at least to anticipate the users' choices. An example is the lighting component and how it can change the perception of an environment simply by acting on the chromatic temperature, or how, by adding an olfactory component that can be repulsive or attractive, it is possible to modify the ways and time of using a space or its path. The same happens with temperature, or the sensory mixture that can take place between the sight of an element that is potentially soft or hard to the touch and that can inspire people to approach it or not. So, it could happen that a multimedia façade like the one designed by d'strict in Seoul, where the power of a virtual wave breaking against the walls of what seems like an enormous aquarium, gets curious before it's even seen.

5.3.9 *Neo-Synesthesia*

The sensory experience generated by a synesthesia can be used in the project as a strategy by taking it also from the semantic point of view, which consists in expressively associating two words relevant to two different personal spheres.

Probably Clino Trini Castelli started from these considerations when in 1992 he designed 'Osmic Gate', or when Diller Scofidio + Renfro thought about a wall of fog for the Blur Building, a pavilion realized for the Expo 2002 in Switzerland (Fig. 5.10).

In the broader sense, projects such as Nendo's Uncovered Skies instalation, a 15 m walkway on the floor of which appear drawings that represent the sky during the four seasons and that are made visible only by using the special umbrellas provided and made of polarized film, such as those that were deliberately removed from the projectors so that they would apparently project only a beam of white light, could also belong to this type of approach. Although it appears that the umbrella's shadow acts like a small screen, in reality, there is one large image projected as if it were 'waiting', visible only as the umbrella moves through space. Or the Wecup project by Studio Squash, designed for large public events with the dual purpose of collecting data and at the same time solving the problem of the large amount of waste that is generated during concerts, festivals, or other large events. The idea behind Wecup is simple: in some areas, two large transparent containers are placed with a sign hanging above each one. By throwing your plastic cup into one of these containers, you vote for the corresponding message.

Fig. 5.10 Blur Building. Swiss Expo 2002, Yverdon-les-Bains, Switzerland. Project By Diller Scofidio + Renfro. Image Courtesy Diller Scofidio + Renfro

5.3.10 *Oxymoronic Landscapes*

As in synaesthesia, the rhetorical figure of the oxymoron involves the juxtaposition of words in the same sentence, in this case, expressing contrary ideas. In the context of the project, the strategy of the oxymoronic landscapes consists therefore in working for a sort of juxtaposition of opposites, according to some possible categories. For instance, you can work on temporal oxymorons, or oxymorons of speed, on a fast or slow journey, or even on the oxymoron of the 'bright night' in which artificial lights are used to modify the perception of space with respect to the day, as in the work 'the sower of stars', a statue located in Kaunas, Lithuania. During the day it seems to represent a simple farmer, a 'fazendeiro' who is sowing, but in the evening the shadow of the statue is projected on a wall and seems to sow dots (the stars), that's the result of anonymous urban art. Even the direction in which a space travels can be an opportunity to create an oxymoronic landscape; working in the back-front sense can allow a surprise effect on the visitor (Fig. 5.11).

Such an example can be considered the installation for the exhibition 'Unbuilding Walls' edited by GRAFT in studio in 2018 for the German Pavilion at Biennale in Venice. Upon entering the pavilion, the visitor was confronted with a large black wall that is seemingly impenetrable; as he got closer, however, he could take the opportunity to discover that this wall was actually composed of many staggered fragments, and each of these, on its white back, told a story, by making the visitor more aware and enriched on his 'return trip' to the exit.

Fig. 5.11 German Pavilion at Biennale Architettura 2018. Project by studio graft. Image Courtesy studio graft

5.3.11 *Leaning to the Context*

When you are dealing with the redevelopment of a space with the left-overs approach, you cannot ignore listening and understanding it, in an attempt to set up a symbiotic relationship between place and project. Hence the strategy of leaning on the context, that is, to use what you already have, what you can find, whether that be materials, stories, or traditions (Fig. 5.12).

The term horror vacui describes a Moorish visual practice adopted by Portuguese builders in the 15th century that involved covering the facades of buildings with azulejos, typical blue and white ceramic tiles that represent historical events.

Taking a cue from this narrative tradition, Jeff Kolb, Ang Li, and Phoebe Springstubb created the installation 'Horror Vacui' for the 2013 Lisbon Architecture Triennale; he effectively combined vernacular production techniques with digital media to temporarily cover the facade of a historic building. The project began with the collection of photographs of interiors, sent by people interested in participating in the project; by using a software, he organized all the material and then he turned it into white and blue colors to be produced and finally mounted on the facade. This

Fig. 5.12 Horror Vacui for the Lisbon Architecture Triennale 2013. Project by Jaffer Kolb, Ang Li and Phoebe Springstubb. Image Courtesy Francisco Nogueira

process produced the final result whereby, viewed from a distance, the tiles were read as azulejos similar to the many others in the city but, getting closer, the facade was understood as a narrating surface, a communicating agent of a guided but collective project.

When we talk about dialoguing with the context, we cannot but mention two masters, one from the past, Carlo Scarpa, and one contemporary, Attilio Stocchi.

When he was commissioned to renovate Palazzo Querini Stampalia in Venice, Carlo Scarpa sought, with his project, an osmotic relationship between the interior spaces and the urban space, and in which water, one of the symbols of the lagoon city, was the protagonist. The high water, in fact, in Scarpa's vision, was not an element to be pushed away, a problem to be solved, but a guest to be welcomed, letting it invade in a planned and careful way the spaces on the ground floor through bulkheads that run along the perimeter walls and create games of reflections and transparencies. Such an extremely refined approach exploits what could be seen as a problem, but which is also a peculiarity of the city, and turns it into poetry.

Go searching for hidden stories to transform them into liveable matter is one of Attilio Stocchi's ways of designing. So, starting from ways, worlds and motions, not always usual for the language of architecture, he manages to stage stories. He does so, for example, with Vortice, a staircase that, built on the arithmetic/geometry of the logarithmic spiral that is inclined by fifty-four degrees and ideally positioned inside the Platonic solids designed by Leonardo, rises from the towpath up to the level of the historical staircase that allows access to the central square of the town of Vaprio D'Adda, in the province of Bergamo; in this place, Leonardo Da Vinci studied and drew the 'vortices' of the river as intertwined hair and flights of birds from the balcony of Villa Melzi, where he was a guest.

A last way of relating to the context is to plan the leftovers with leftovers, as happened for the Geopark, a playful urban space on the waterfront of Stavanger, Norway, a famous site for offshore oil bases. The surfaces and installations, in fact, were created from recycled elements of oil rigs, the abandoned Frigg oil platform, offshore bases, equipment suppliers and scrap piles. Thus, what was once an urban void is now a gathering place at any hour.

5.3.12 Close-Up

As mentioned in the previous paragraph, one of the most interesting approaches to the project of Avanzi is to identify the main features, peculiarities, potentialities, even if they are simple details, so as to emphasize and tell them from new, peculiar, or special points of view, maybe through close-ups. Close-Up is a typical technique of cinema and photography, which allows you to look closer at someone or something than the real point of observation. Through the Close-Up an alternative design strategy evolves it means thinking about a condition straight from inside, it means being involved, paying attention to small things, even with the risk (or potentiality) of producing disorientation, of canceling the distance between 'minor instances' and

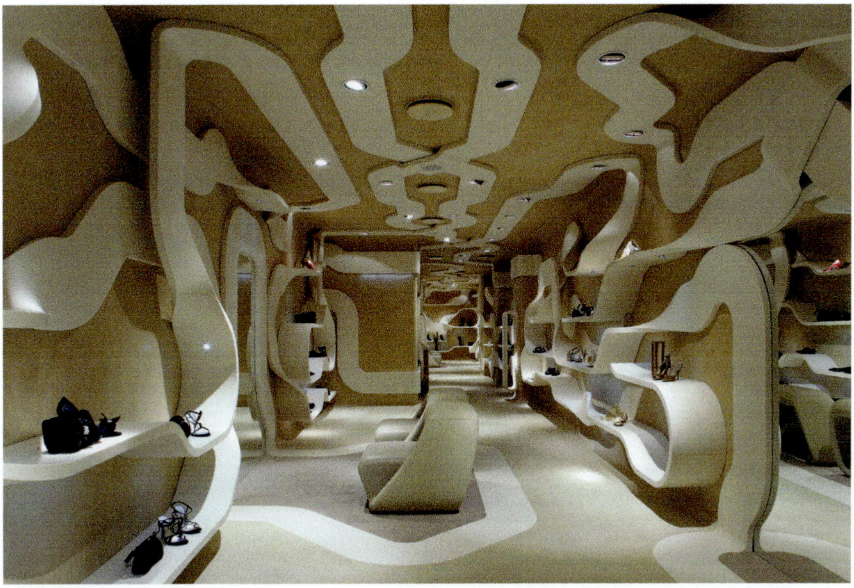

Fig. 5.13 Stuart Weitzman Store, Roma, 2013. Project by Studio Fabio Novembre Image Courtesy Studio Fabio Novembre

'major instances' and subverting a contemporary and global project approach that tends to cancel details (Heindl 2013) (Fig. 5.13).

You can think of working by synecdoche, the rhetorical figure that identifies, among the various meanings, the part for the whole; and perhaps Fabio Novembre must have had this principle in mind when he used the string, enlarging it in scale, to create all the components of the Weitzman shoe store in Rome, from the shelves to the lights, to the seats up to the floor graphics. String is also the protagonist of Marko Brajovic's project for the Camper Stores he designs: a multitude of red strings, the brand's color, that are hung from the ceilings, creating ever-changing sinuous shapes inspired by the places where the stores are located, following the 'think global, act local' approach. The theme of the repetition of detail is not new to the world of Camper, so much so that from the very first stores, designed by Martì Guixé, shoe boxes were used as an element of repetition to create the aesthetics of the space, or the store in New York designed by Nendo in which white silhouettes of shoes alternate with real models that are then made immediately recognizable by the chromatic cancellation of the surrounding environment (Fig. 5.14).

Repetition is clearly a quick and successful way to approach the design of a space, as I experienced firsthand with the setup for the launch of the new colors of Arjowiggins' papers. The 70 × 100 sheets with all the new colors were hung from the ceiling to create a chromatic cascade inside the space, a photo studio in downtown Milan with skylights that can be opened and darkened; this feature was exploited to create vibrant effects of light and movement at certain times of the day.

Fig. 5.14 Colours! By Arjowiggins, Milano, 2018. Project by emiliolonardo^{design}. Image Courtesy Sara Mattaioli

And reasoning by close-up can also give life to poetic projects for often neglected objects (Fig. 5.15).

This is the case of the graphics designed by Matteo Ragni and Giulio Iacchetti for the Montini manhole cover company, which distance themselves from a purely decorative operation in favor of a narrative process in which the manhole cover dialogues with the surrounding environment, thanks to its skin.

5.3.13 I'll Spite You

One of the principles of the architectural restoration as stated in the Athens Charter is to respect the work of the past. Working, however, by adding irony to one's projects, can foster a new perception of the advanced space; hence the strategy of 'spite you'. This strategy allows us to think and try to solve social problems which are raised by the project. As it happened to 'Out in Prison' by Gabu Heindl, who won the competition to redesign the outdoor courtyard for men's recreation at Krems prison, in Lower Austria. The project, that is an unlikely soccer field made of synthetic grass, succeeds, with its realization, in making evident a problematic issue: despite the small and insufficient dimensions of the space to accommodate a regulation soccer field, Heindl, lowered the design of the field from above, by using the behavior of cut & paste, and forced some of its parts to fold on the perimeter walls. And it is exactly

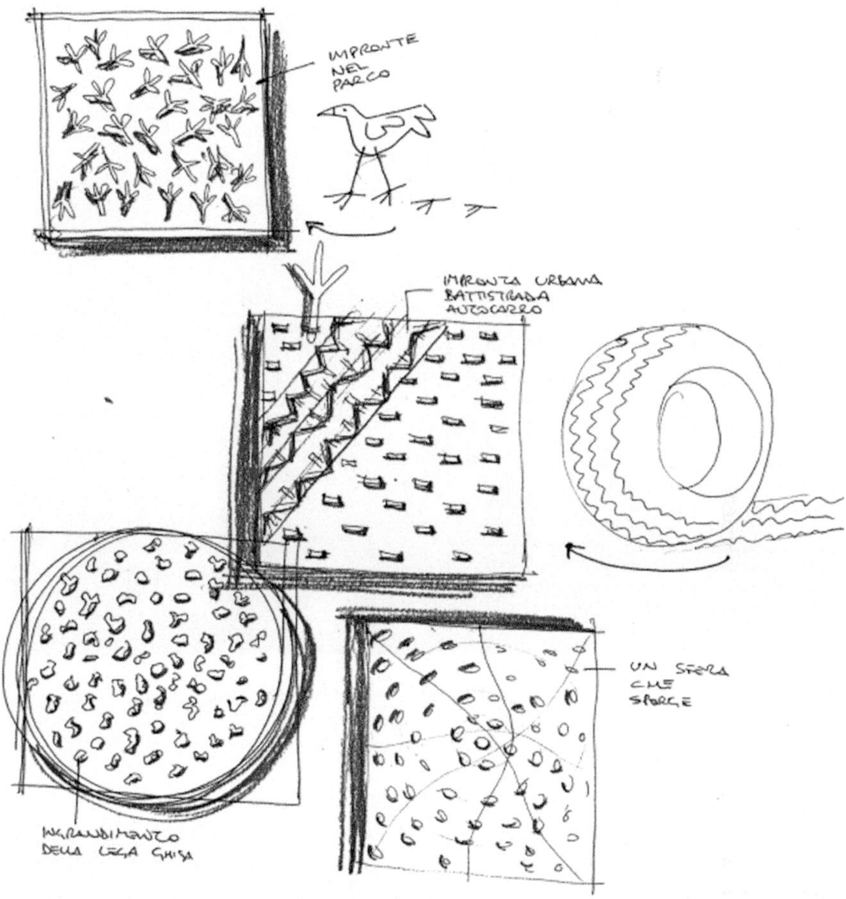

Fig. 5.15 Tombini. Project by Giulio Iacchetti and Matteo Ragni for the brand Montini, 2012. Image Courtesy Giulio Iacchetti

this lack of space, without filters and comments, to communicate the essence and the physical impact of a restrictive condition that must be modified (Foppiano 2013) (Figs. 5.16 and 5.17).

The spite strategy can also involve other design attitudes, such as that of confusing; Nendo set up the Japanese design exhibition at the National Design Center in Singapore in 2014 by using isometric graphics to create an alienating effect of three-dimensional and two-dimensional objects that play at altering the visitor's perception. In a similar way, and to emphasize an abandoned lot—a leftover—in the city of London, Argentinian artist Leandro Erlich in 2013 staged the work Dalston House. The work, which resembles a film set, reconstructs the full-scale facade of a late nineteenth-century Victorian townhouse. The façade has been laid on the ground, while a mirror surface tilted at a 45° angle is positioned above it. The audience, as

Fig. 5.16 Outside in Prison, Krems, Austria, 2011. Project by Gabu Heindl. Image Courtesy Gabu Heindl

in many of Erlich's works, plays an active role in bringing the installation to life, as while remaining seated, standing, or lying on the horizontal plane; the visitors appear to be intent on scaling the building or appear to be hanging outside the building. A further approach to the strategy of spite can parasitize the space in order to modify its perception or the way it is used. In this way, certainly some of Numen/For Use's instalations are an example: with the 'Tape' series, the Croatian collective creates site-specific installations using adhesive tape. The spaces are thus parasitized by huge artificial organisms similar to cocoons, some of which can be walked inside, usually tracing the usual pathways of the spaces and forcing visitors to experience and look at the places in a different way than usual.

5.3.14 Strategies in the Final Synthesis Studio

Since the academic year 2015/2016 in the Studio of Final Synthesis of the School of Design at Politecnico di Milano, carried out by Luciano Crespi, we have been working on the issue of Leftovers. After a first analysis on the theme, we ask the students to elaborate a Design Metaphor, that is, a concept, their original guideline

Fig. 5.17 Hidden—Unveiling Japanese Design, Singapore, 2014. Project by Nendo. Image Courtesy Masaya Yoshimura

throughout the semester, a compass they can use to guide themselves in making decisions during the succession of the project phases; and as cardinal points, the strategies can only be daughters of the metaphor previously elaborated. So, for example, metaphors such as 'Echo' are resolved through strategies that include reverberation, distortion or repetition; 'Paradoxa' with the inversion of the rules, the out-of-scale, the optical illusion and decontextualization. Some are definitely more successful than others, but the balance of these years has led us to underline some clusters of approach. In fact, in the students' projects we can recognize recursive ways using peculiar languages, which have frequently been borrowed from other disciplines. In particular, we have metaphors that draw on the language of cinema/theater, such as 'Cliffangher', 'Strangulation', 'Messa in Scena' or 'Fuori Campo', which therefore used strategies typical of these fields, such as detachment, distortion of the everyday, or emphasizing the mechanism; on the contrary, others words have been borrowed from the literary/musical world, using approaches typical of the music language or other rhetorical figures as Metaphors, as in the case of 'Enjambment', 'Climax', 'Aposiopesis', 'Oxymoron', 'Hyperbole' or 'Dissonance', with techniques as emphasis, veiling (or unveiling), contrast. Another area from which students often took inspiration is the broader scientific field; thus, metaphors as 'Contamination', 'Pulsar' and 'Multiverse' and also strategies like hybridization or recombination have been used.

5.4 Learning to Observe

At the end of this long text, I would like saying, once again, that one of the most important characteristics that the designer must train, and now we return to the opening lines of this chapter as a road that takes us back home, is the ability to observe, to grasp, even in the apparently insignificant, a mine of possibilities. It is only in this way that a simple wall can become a living space, such as in the project 'Intra Muros' by the Chapuisat brothers, or a Tea Bar, as in the Alan Chu's project in Sao Paulo, or even, going even further back in time, a functional facade such as the renovation of the Storefront for Art and Architecture in New York that Steven Holl and Vito Acconci rethink as a device to unite where it usually divides, symbolically representing Art itself. This is what we try to convey to our students with the method that we repete and refine year after year, also thanks to them. And if we want to elect an item as a metaphor for all this, I can only choose the bench, that, as Ugo La Pietra suggests, will not only be a place for a pleasant rest but also an observatory from which it is possible to look, to glimpse, to contemplate (Fig. 5.18).

So, I hope to the reader that, after reading these pages and understanding the value of Leftovers, he may start looking at the world with different eyes, maybe with the eyes of a child, who is able to invent new worlds with his imagination.

Enjoy your journey!

Fig. 5.18 Intra Muros #1, Basel, Switzerland, 2006. Project by Gregory and Cyril Chapuisat. Image Courtesy Gregory Chapuisat

References

Botti S (2017) Editoriale. Abitare 568:15
Ciuffi V (2012) Grindbakken. Abitare 527:84
Crespi L (2013) Da spazio nasce spazio. L'interior design nella trasformazione degli ambienti contemporanei. Postmedia Books, Milano
Croce B (1992) Storia del Regno di Napoli. Adelphi, Milano
Debord G (1955) Introduction á une critique de la géographie urbaine. Les Lévres Nues 6:288
Foppiano A (2013) Out in prison. Abitare 534:37
Heindl G (2013) Close-up. Abitare 533:19
Lefebvre H (1976) Il diritto alla città. Marsilio, Padova
Mari E (2011) 25 modi per piantare un chiodo. Mondadori, Milano
Pallasmaa J (1996) The eyes of the skin: Architecture and the senses Polemics. Academy Editions, London
Pine BJ II, Gilmore JH (2000) L'economia delle esperienze. Etas, Milano
Poli C (2009) Città Flessibili. Intar Libri, Torino
Rovelli C (2017) L'ordine del tempo. Adelphi, Milano

Part II
International Scenario

Chapter 6
Revisiting Heritage and Regenerating the City: The Rest as a Resource in Gulf of Benin

Monica Coralli and Didier Houénoudé

Abstract This chapter examines the urban transformations underway in South Benin through the prism of heritage. As the text is organized into six points which structure it, each of them develops specific approaches to the heritage dimension. Illustrated by one or several significant examples, the realities or projects described offer answers, never clear-cut, on rehabilitation or, on the contrary, on the occultation of the past. Thus, while events, by definition ephemeral, have difficulty in reviving the colonial heritage in a different way and giving it meaning, initiatives led by artists' collectives periodically attempt to stimulate urban regeneration dynamics. Their mixed results over time also demonstrate the weaknesses of national policies which do not yet manage to propose actions on several scales and whose articulation is designed to activate dynamics likely to sustainably revitalize certain urban realities.

Keywords Wasteland · Heritage · Patrimonialization · Urban regeneration · Benin

In the large African metropolises, built-up areas such as open, public, wasteland spaces, neglected for a more or less long period of time by the public authorities, almost systematically call for informal, illegal but generally well-tolerated or even encouraged occupations, as long as a new development or construction project does not invest in them. Whether they present remarkable architectural or landscape characteristics, or on the contrary have no singular value, these spaces are very often used as squats, or as warehouses, or they become workplaces, open to various and cumulative uses, even though a priori incompatible. The Hunkanrin house in Porto-Novo (Benin) is one of these (see next page).

It also happens that places recognized as UNESCO World Heritage Sites or likely to be inscribed there (built and unbuilt areas), and which have therefore already received special attention from the authorities (communal, state, etc.), are neglected

---

M. Coralli (✉)
Laboratoire Architecture Anthropologie, CNRS, UMR 7218, Lavue, Paris, France

D. Houénoudé
National Institute for Art, Archaeology and Culture, University of Abomey-Calavi, Abomey-Calavi, Benin

L. Crespi (ed.), *Design of the Unfinished*, The Urban Book Series,
https://doi.org/10.1007/978-3-030-73457-2_6

and even the facilities made for their development are quickly damaged, due to lack of maintenance (cf. Heritage House and Mosque of Porto-Novo, Governor's House in Ouidah, etc.). What are the reasons for this paradox? We formulate the hypothesis that these are to be sought in the gap between the recommendations made by UNESCO, whose current legislation is reflected in Benin as in many other countries of French-speaking Africa, and what the communities—and each individual—consider to be "their heritage" Coralli, Houenoude (2013). Indeed, the modalities of construction of the heritage value of an object or a site differ from one case to another, according to their nature or status, or even according to the attachment that they arouse.

Thus, our work has consisted in deciphering the heritage dynamics underway in a few cities located on the coast of the Gulf of Benin, in particular, Cotonou and Porto-Novo in the Republic of Benin, through a few examples of spaces or buildings whose transformation phases or, where appropriate, project ideas have followed one another. Thus, our reflection was based on studies of possible developments or direct observation of actual developments: some of them could give new life to certain spaces in these cities, while others have already extended the life of the latter. Both are trying either to rediscover old ways of living there, to reinforce those that have developed over time or, finally, to invent new ones.[1]

6.1 The Hunkanrin House or to Occupy While Waiting

Located near the new building of the National Assembly, in the former neighborhood of "foreigners", especially Afro-Brazilians,[2] this house is a former residence of Afro-Brazilian type[3] on one level. The building has five rooms including three bedrooms, a living room, a dining room and a terrace at the back of the house. It has a backyard in which the well, the toilet and the ruins of the kitchen and a printing shop are

[1]This text extends the reflection we have been jointly carrying out for the past 10 years on heritage in Benin. After the article published in *Espaces et Sociétés* in 2013 and several communications on the subject, we have opted here for a focused look at specific examples which illustrate, at all scales, the slow implementation of an effective heritage policy in the Beninese context. Artistic interventions find a more important place here. They do not manage to generate the hoped-for dynamics of urban regeneration.

[2]The Afro-Brazilian community that was formed between the eighteenth and nineteenth centuries is in fact a composite group of freed slaves or slaves who had bought back their freedom and returned from Brazil, Portuguese merchants who had been living on the West African coast for longer periods of time, and servants of the two previous groups who adopted the Europeanized names and ways of doing things of their masters.

[3]The great mosque of Porto-Novo, by its construction materials, mainly mud brick, and its style, is one of the visible traces of the influence of the Aguda in Benin who, moreover, knew how to define a very particular style for the construction of their dwellings. Different from the low Portuguese Sant-Louisian houses or other variations of the Portuguese style imported to Brazil and returned via the returning migrants, they are a clear sign that the journey from America to Africa led to diverse local interpretations of this same style. Moreover, all along the Gulf Coast of Guinea, similarities can be observed, particularly in the decoration of their facades, the dimensions of the bays, the tiled roof and the organisation of the interior spaces.

Fig. 6.1 The Hunkanrin-Faihoun House. The building is left abandoned and is now occupied by mechanics. Photo: M. Coralli, 2020

located. The entrance garden has been reduced by the construction of two stores and a fence wall (Figs. 6.1, 6.2, 6.3, 6.4 and 6.5).

This building has existed since the end of the nineteenth century (it was built around 1890). The builder is a German merchant named Hunger Bauer who built it for his daughter Suzanna Hunger Bauer as a wedding dowry. She died in 1934.

The building is bought by Louis Hunkanrin under the cover of the name of his friend Faihoun, for fear of confiscation of his property by the colonial administration. Indeed, Louis Hunkanrin was a teacher of his state and an anti-colonial activist. He created a newspaper, Le Messager dahoméen,[4] to convey and carry criticism of the colonial administration. An object of numerous threats, he was finally exiled in Mauritania for 10 years. On his return to Dahomey, he wrote in *La Voix du Dahomey*[5]

[4]*Le Messager dahoméen* is a newspaper created by Louis Hunkanrin in Paris in November 1920, 2 months before his repatriation from France in January 1921. The political direction of the newspaper was entrusted to a lawyer at the Paris Court of Appeal, Mr Max Clainville-Bloncourt. In the newspaper, Hunkanrin denounces the abuses and inconsistencies of colonialist policy in a vitriolic manner.

[5]*La Voix du Dahomey* is an anti-colonial newspaper strongly influenced by Blaise Kuassi, an admirer and son-in-law of Hunkanrin. The latter collaborated actively in the publication of the newspaper, which was facing a lawsuit brought by the colonial administration from January 29 to April 9, 1936.

Fig. 6.2 Entrance to the abandoned Hunkanrin-Faihoun house and yet listed as if it were occupied by the addressing services of the City of Porto-Novo. Photo: M. Coralli, 2020

and was fined after the trial against the newspaper in May 1936. During the Second World War, Hunkanrin was accused of "Gaullism" and deported to French Sudan. Returning to Dahomey after the war; he no longer participated in political life and lived in the building until his death, after which his daughter took over the house.

A conflict of inheritance then opposes his daughter to the Faihoun family following Hunkanrin's death. Despite the possession of the original deed of sale and tax receipts by the daughter, the court ruled in favor of Faihoun's heirs.

At the death of Louis Hunkanrin's daughter, the house remains unoccupied and falls into ruin.

The site has since become a squat occupied by motorcycle mechanics who also store spare parts for two-wheeled vehicles. These mechanics sometimes dismantle the building for their own use, destroying much of the richly carved woodwork that they have regularly used to make fire.

It all started with the publication, in September 1934, of a story denouncing a case of corruption of a high-ranking civil servant. The affair caused the administration to be in turmoil, and the newspaper's premises and the manager's home were searched. The newspaper's admirers, including Hunkanrin, were arrested. At the end of the trial, 28 fines were imposed for "illegal formation of association", two for "theft and concealment of documents", and ten for "breach of respect due to French authority". Among those convicted was Louis Hunkanrin, who was fined 100 francs.

Fig. 6.3 The ruined Hunkanrin-Faïhoun house is now a threat to those who frequent it. Photo: M. Coralli, 2020

A building renovation project thought up by two citizens of the city of Porto-Novo planned to give the building a second life. The objective of the project was to turn the abandoned space into a center dedicated to the conservation and valorization of old photography.[6]

The new destination of the building thus made it a kind of photographic archive conservatory, with a space dedicated to digital databases and another space for temporary exhibitions. Another aspect of the project was to set up a press area in memory of Louis Hunkanrin.

The renovation project, although funded and supported by the municipality of Porto-Novo, is faced with the refusal of the Faïhoun heirs who hardly ever resided in the house and who seem to prefer an illegal squat that would progressively destroy the building instead of a restoration that would restore its value and rehabilitate the memory of Hunkanrin, the real owner of the premises.

[6]This project echoes a few house rehabilitation projects on the island of Saint-Louis in Senegal where a patron of the arts, Amadou Diaw, from Saint-Louis, has acquired about ten houses and transformed three of them into museums and cultural venues open to the Senegalese and international art scene. In particular, the Museum of Photography (Mupho), inaugurated in November 2017, which has set itself the mission of becoming a photographic archive centre, bears witness to the rich photographic tradition of Senegal and the sub-region and recalls the precursors and masters of photography since the nineteenth century, then the golden age of studios up to the present day. Its initiator entrusted Salimata Diop with the artistic direction.

Fig. 6.4 Much of the Hunkanrin Faihoun house is severely damaged. Photo: M. Coralli, 2020

The Hunkanrin-Faïhoun building presents a remarkable architectural style in an environment that fits into the orthogonal framework of the colonial period. Nevertheless, the illegal occupation of the plot by several protagonists has caused a degradation of the building, which has also led to a reduction in the surface area of the site.

Nevertheless, it is still possible to rehabilitate the building by modifying its use if the family's concerns are settled by an agreement with the local authorities, who could contribute or even take charge of all the necessary work.

6.2 The Past Between the Allochthonous Will Be to Rehabilitate and the Local Desire to Erase It

The colonial project and the modern project in its various facets, coupled with a political ideology and a new idea of society, have produced spaces that are today neglected and left behind. If some of these spaces have survived or even had another life, it is above all because the administration inherited them when the country acceded to national and international sovereignty. Most of these colonial buildings were invested by the public services, which gave them a new purpose. This is the case

Fig. 6.5 The Hunkanrin-Faïhoun building is in a state of abandonment, even though it is located in a district where the construction of the new Beninese parliament was planned, and has since been halted. Photo M. Coralli, 2020

of the Porto-Novo Post Office Directorate and the Governor's Palace, which has now become the Beninese Parliament. Other buildings, on the other hand, enjoyed other fortunes. Indeed, some buildings, because of the mixed or painful memories they evoke, have been mostly left abandoned, as if to throw the veil on a painful past that one wishes to erase from the memories. In this respect, we can cite the residence of the Commander of the Circle of Ouidah (Fig. 6.6). However, this residence was the subject of a guest house project designed by Sandra Dossou and Richard Sogan (2007), but unfortunately the project did not materialize.

The building, currently disused and in a very advanced state of degradation, was built by the colonial administration in 1890. It served as the residence of the Commander of the Circle and at the time housed his quarters on the first floor, rooms

Fig. 6.6 Former residence of the commander of the Cercle de Ouidah. Photo: D. Houénoudé and R. Kessou, 2013

for distinguished guests on the second floor and offices on the ground floor. The residence underwent numerous modifications which altered part of the architectural homogeneity of the building.

An enhancement project has been set up by some Beninese executives, encouraged by French cooperation. The project did not prosper because some of the town's inhabitants saw the support provided by French cooperation as a way of celebrating colonial France in the town of Ouidah.

However, the building is now one of the emblematic constructions of the city of Ouidah. It is erected on a site located at the back of Ouidah's Town Hall. The building is badly deteriorated, but it remains the object of many covetous desires both on the part of the various tourist actors, the heads of establishments and the municipality itself, which owns it.

Although the latter has had the town hall entirely rebuilt on the model of the Slave's House in Island of Gorée, Senegal, (double staircase) reminiscent of a colonial building, the House of the Commander of the Circle was not included in any restoration project for offices in the municipality, which tends to feel cramped despite its new site. The latter has nevertheless been spotted by the heads of the University of Abomey-Calavi who are planning to make it the site of their School of Hotel Management and Tourism, it being understood that Ouidah figures prominently among the cities with high tourist value. Thus, without waiting to obtain all the approvals and authorizations, the managers of this school in the process of being created have already erected a plaque on the site to indicate the appropriation of the space by them.

6.3 Urban Initiatives Financed and Supported by Public Institutions

Many of the buildings in very poor conditions were mostly redesigned by groups of people sensitive to the history of the buildings. These are often projects carried out by associations or individuals requesting, sometimes, with the support of political-administrative bodies such as municipalities (Diaspora Media Library at the Place des Martyrs in Cotonou). However, it must be noted that, due to the difficulties posed by the ownership of the property, few of these projects are successful. It is easy to understand the hesitations when it is a question of investing in a property which the owner can claim at any moment. Houénoudé (2013a, b) describes how problems of joint ownership can jeopardize building restoration projects. Thus, the types of occupation or diversion of use or even development of these vacant spaces were rather sketchy, unless a rehabilitation lease project was concluded, as is the case for projects that have been able to be carried out.

Four projects of this type could be carried out in Porto-Novo and these projects obtained the consent and support of the municipality, even if the latter is considering putting the last three under private management. One of the first and which served as a model for the others was the Maison du Patrimoine et du Tourisme (Heritage and Tourism House).

The building, built in 1912, now belongs to the Migan Sohè family. For a long time abandoned because each member of the family had built his own house, it was identified by heritage professionals to house the headquarters of the Maison du patrimoine et du tourisme of Porto-Novo. The choice of the house took into account its privileged location in the heart of the big market of Porto-Novo. The idea was that the example of the building's rehabilitation could be emulated and push the owners of buildings in ruins to opt for renovations that could restore the value of their property.

The project was financed by the cities of Lyon, Cergy-Pontoise and Porto-Novo, and the restoration was carried out in the form of a worksite school to familiarize young people and workers with the techniques of restoring old buildings.

The restored building now houses a city department dedicated to culture and cultural heritage (Fig. 6.7).

The restoration of the three other buildings is financed by the World Bank through the PAURAD project.[7] The renovated houses will be managed by a private company

[7] The Urban Development and Decentralisation Support Project (PAURAD) is a project implemented under the supervision of the Ministère du Cadre de Vie et du Développement Durable in collaboration with the World Bank. Its objectives include architectural and technical studies, control, and supervision of the rehabilitation works of the three Afro-Brazilian houses in the city of Porto-Novo.

Fig. 6.7 Maison du Patrimoine et du Tourisme (former Migan house restored in 2005–2006). Photo: D. Houénoudé, 2013

under contract with the municipality and will serve as guest houses with a restaurant and bar for some of them (Ministere du Cadre de vie et du Developpement Durable, 2018a, b, c).

In some cases, buildings have been identified by school principals who express the wish to invest in them in order to establish their schools, which are generally interested in cultural issues (heritage, architectural and/or tourism). This is the case of the CFAO building located on the edge of the lagoon, a few steps away from the new-and-old National Assembly, the erection of which has been suspended. It is made entirely of stone with interior wooden structures that are almost completely destroyed. The building, however, is inhabited by a squatter family, who view with displeasure any attempt to take away what they consider to be their home. The ownership of the building is indeed ambiguous. The municipality of Porto-Novo and the ministry in charge of finance are fighting over some of the buildings inherited from the colonial period (Fig. 6.8).

The Institut National des Métiers d'Art, d'Archéologie (INMAAC) approached the town council, targeting the CFAO building and another private building for which the town was in negotiations with the owners to restore it and make it a center for studies in the restoration of the built heritage. The project to transform the latter

The activities of the PAURAD are divided into three components. Component A concerns, among others, the rehabilitation of cultural heritage (royal palaces, Afro-Brazilian houses), etc.

Fig. 6.8 Warehouse of the CFAO built around 1910, today abandoned by the public authorities and squatted by unfortunate inhabitants who have taken up residence there. Photo: D. Houénoudé, 2020

building into a university establishment went quite far through the Municipality's request for a renovation plan drawn up not only by an architect, but also by incorporating student reflections on potential developments and restorations of the building. The project, although having obtained the agreement in principle of the previous mayor, could not get off the ground due to a lack of resources and firm support from the municipality.

The same is true for many projects carried out by third parties which are struggling to prosper due to a lack of means, but even when these means exist, the hesitation of project leaders faced with a risky investment sometimes takes over. Indeed, the risks that an investment in a development, even a temporary one, raises are likely to discourage the most daring who fear that, in the absence of a contract signed in due form with the owners, the latter might take over the project, destroy it or use it to enhance their own development (Figs. 6.9, 6.10, 6.11, 6.12, 6.13 and 6.14).

Within the framework of a course module for the Bachelor of Architecture degree at the National University of Science, Technology, Engineering and Mathematics

Fig. 6.9 The Padonou Aminou House. Photo: D. Houénoudé, 2012

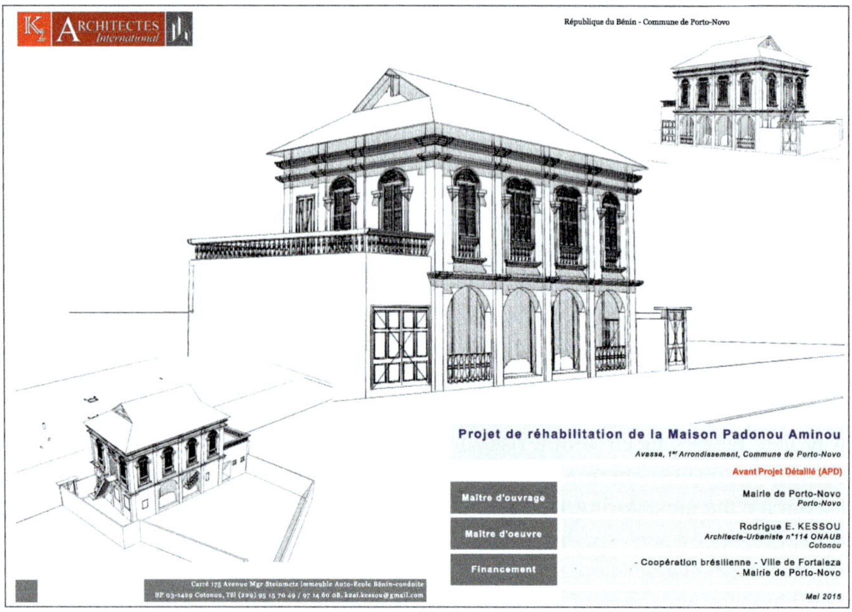

Fig. 6.10 Rehabilitation project of the Padonou Aminou House, Arch. R. Kessou, 2015. Student screenings on the possible rehabilitation of the Maison Padonou Aminou

Fig. 6.11 *Author* Hevou Isaac, 2015

Fig. 6.12 *Author* Eklou Johnson Samir, 2015

(UNSTIM) in Abomey, we asked the students as an exercise to reflect on the rehabilitation of the Padonou Aminou house as a training center for the restoration of architectural heritage. The idea was to take the ruined building as a starting point

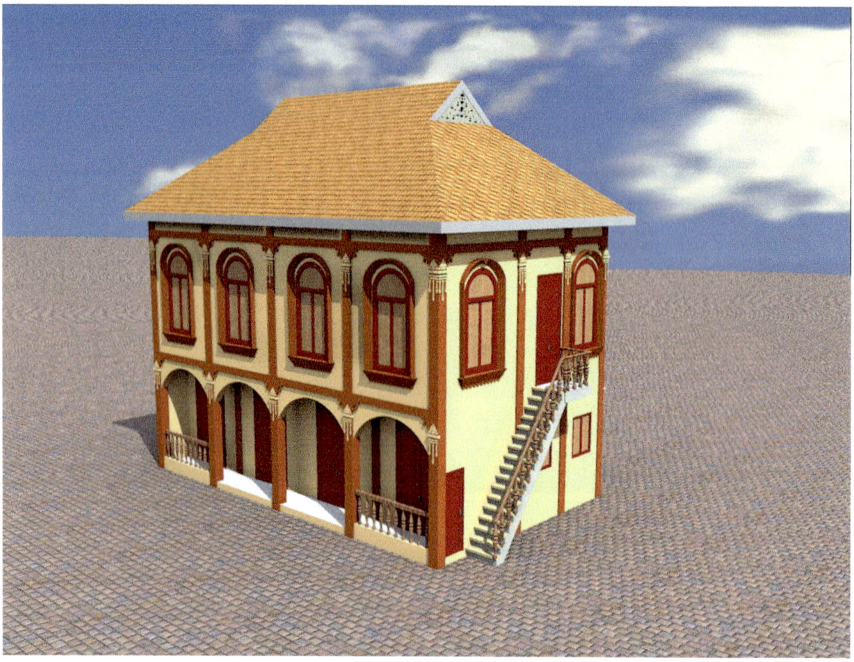

Fig. 6.13 *Author* Akinocho Alabi, 2015

Fig. 6.14 *Author* Yessoufou Saïd, 2015

for a virtual reconstruction and to give it colors that could symbolize its renewal and refer to the characteristics of Afro-Brazilian houses.

6.4 Events as a Means of Reactivating Places: Giving Meaning to "Forgotten" Sites

On the occasion of the jubilee of Benin's accession to independence, the Beninese visual artist Edwige Aplogan decided to cover with the flags of the African countries that gained independence in 1960, the most emblematic monuments of the country and which have an important political history. For her, "to cover national historic monuments is to remember our martyrs who fought for us to achieve independence and at the same time to question this facade of independence that some of us deplore". In Porto-Novo, Edwige Aplogan was interested in draping the city's first four-storey building belonging to former President Sourou Migan Apithy, one of the first civil presidents from the south of the country. Located in the Afro-Brazilian quarter, the building was called "Fort Ouidah", in analogy with its massive shape and location in the Afro-Brazilian fabric. The building has been abandoned for many years and is rumored to be haunted.

Apart from the drapery made by Edwige Aplogan, almost no squatting or intervention activity can be observed in and on the building, although most of Sourou Migan Apithy's beneficiaries hardly reside in Benin anymore, and their absence could therefore have facilitated temporary appropriation of the space by third parties.

The building's reputation as a haunted building, rather than the fact that it belonged to one of the country's first presidents, would certainly play an important role in this lack of temporary appropriation. The same applies to many abandoned spaces or houses that are often not invested because of the presence on the premises of vodun altars belonging to the owner's family (Figs. 6.15 and 6.16).

Other spaces neglected by the state in its development projects, although they enjoy a highly symbolic place in the history and memory of the Beninese (Adegbidi et al., 2012), continue to "exist" because of occasional interventions or the tolerated presence of private actors on the site. This is the case, for example, of the Place des Martyrs, renamed Place du Souvenir by Nicéphore Soglo, the first president of the Democratic Renewal in the 1990s.

The Place des Martyrs was erected to commemorate the armed aerial aggression against the People's Republic of Benin during Operation Shrimp by a group of mercenaries led by Robert Denard[8] on January 16, 1977. The monument was built with the support of the Republic of Korea (North Korea) to pay tribute to the seven Beninese who died in the attack.

The monument, which has offices in the basement, has been taken over on the ground floor by an association which has housed its offices there and which, on an

[8]Robert Denard, known as Bob Denard, is a French mercenary involved in numerous coups in Africa.

Fig. 6.15 Apithy building, owned by a former president of the Republic of Dahomey. Photo: C. P. Tossou, 2010

occasional basis, organizes exhibitions of works of contemporary art and a semblance of a museum of the Beninese Armed Forces. The whole thing is now in a relative state of abandon. The walls have deep cracks, while the ceiling allows water to filter through. The occasional occupants, who have come to be considered by the inhabitants as part of the décor, try as best as they can to bring the space to life alongside a bus-restaurant, installed on the square's esplanade, which is very popular with Beninese people who work in the surrounding services, notably at the ORTB,[9] in the Ministries and other private agencies nearby.

From time to time, the esplanade lends itself to occasional cultural events. Thus, for a few days, the site can host stands selling various objects, fashion shows or, more often than not, exhibitions or, finally, concerts and theater and dance performances. The International Theatre Festival of Benin (Fitheb), considered as the most important theater festival in Africa, which should have held its 15th edition in 2020, and which over the years has become an unmissable event for programmers, presenters and professionals of theater and performing arts from Africa and the world, chooses this space as the focal point of the main events organized at each edition (Figs. 6.17 and 6.18).

[9]Office de Radiodiffusion et Télévision du Bénin.

Fig. 6.16 Apithy building draped by Edwige Aplogan on the occasion of the jubilee of Benin's independence. Photo: C. P. Tossou, 2010

At the Place des Martyrs, the link between art and the city is more than elsewhere obvious: it is undoubtedly a high place for events in the perception and representation of the people of Cotton, who gladly invest it by taking part in the events that take place there. Its networking with other venues, regularly included in the Fitheb's program and other regular cultural activities, makes this public space a highly appropriate anchorage point for inhabitants of all ages, less for the memory of the history mentioned above and condensed in the statue than for the succession of activities that are planned there and the constant presence of a snack bar that enlivens the place at all times.

6.5 Spaces of Memory and Spaces of Identity

The monument of the Gate of Return was built in Cotonou, behind the Presidency of the Republic, echoing that of the Gate of No Return in Ouidah. The monument remained unfinished for many years. In 2012, within the framework of the *Biennale Regard Bénin*, an association of visual artists decided to dress it provisionally, in a logic that was intended to challenge the power in place for a definitive objective.

Fig. 6.17 Exhibition "Mino" which was launched on August 1, 2018, on the Danxomè Amazons, elite warriors of the army of the Danxomè kingdom. The exhibition was organized under the patronage of the Ministry of Culture and the Claudine Talon Foundation, the First Lady's foundation. Below is the poster. Photo by D. Tossou (2018)

The project entitled *La porte de retour en couleur* (the door of return in color) envisaged a refurbishment and an almost complete finishing of the unfinished monument. For the members of the association, the aim was to create a "trompe-l'oeil" decorative covering and to deploy a minimum of amenities on the site to allow "the Beninese people to visit it and to see it as a major work (of the) country".

The arrangements planned by the association provided for the various aspects of the monument to be dressed to give the impression of a finished monument, without touching the very essence of the work. Thus, as with all ephemeral installations, this door had to be uninstalled in order to allow the raw cement structure to be restored to its original appearance.

The structure is now destroyed. The Beninese government of Patrice Talon, who decided to build a triumphal way at the back of the Presidential Palace and which joins the international airport, preferred to demolish the monument, erasing or rewriting the history of the memory linked to the transatlantic trade by preferring to erase the statue of an Amazon who can be seen today behind a thin green fence surrounding the scaffolding of the building site.

The Return Gate was a monumental, architectural and artistic work composed of a 14-meter Grand Entrance bearing at each end a set of seven pillars of different heights, symbolizing the crowd that came to welcome the brothers of the diaspora.

Fig. 6.18 Poster of the exhibition "Mino" organized by Michèle Wêni

At the extremities of the Gate stood two giant braided double-sided palms, 14 meters high. They gave a noble character and sacred value to the Entrance.

On the seaward side of the Entrance were two giant sculptures of women, one pouring water to welcome people to Peace and the other offering water to drink to quench the thirst of foreigners who had come from far away.

The side of the Entrance overlooking the City bore two large sculptures of musicians, one playing a talking drum and the other blowing a horn to announce and celebrate the reconciliation between brothers separated by history in time and space.

Behind each sculpture was a large hand, five meters high, symbolizing openness, welcome and hospitality.

On the side of the Entrance facing the sea, large panels were lined up, decreasing on each side, on which frescoes of various structures, textures, materials and themes were to be created to magnify the richness and beauty of Beninese social, cultural and artistic life.

On the side of the Entrance overlooking the city, large panels were also lined up in the same way, on which frescoes of various structures, textures, materials and themes were to be painted to evoke the history and the need for this Reconciliation.

Two large bands called "The Band of Hope" overlooked the structure and intertwined at 23 meters above the Gate and culminating at 28 m.

The project, led by the Association for the Promotion of Young Creation in the Plastic Arts (APJ-CAP), was very specific about the renovation of the monument. Each part of the monument was the subject of a detailed, three-dimensional design project (Figs. 6.19, 6.20, 6.21, 6.22, 6.23, 6.24 and 6.25).

The project of the Beninese artists on the return door did not intend to invest in a sustainable way the space which had been abandoned by the Beninese state. They wanted to make a big splash that would push the State to rethink the finishing touches of the work. However, the artists' collective did not obtain the necessary authorizations to use the space of the monument.

Fig. 6.19 Gate of Return (today destroyed). Photos: APJ-CAP, 2012

Fig. 6.20 Detail of the Gate of Return. Photos: APJ-CAP, 2012

Fig. 6.21 Artistic projects planned by the Association for the promotion of young creation in plastic arts within the framework of the Biennale Regard Bénin. Photos: APJ-CAP, 2012

Fig. 6.22 A disaffected wing of the return gate monument. Photos: APJ-CAP, 2012

Fig. 6.23 The same wing of the monument reimagined by Beninese artists. Photos: APJ-CAP, 2012

Fig. 6.24 The second wing of the monument of return in a state of degradation. Photos: APJ-CAP, 2012

Fig. 6.25 The reconversion project of the gate of return imagined by the artists. Photos: APJ-CAP, 2012

6.6 From Intervention on the Isolated Object to the Systemic Logic of the Regenerative Project?

As we have seen, spaces left fallow can be the object of temporary squatting, as long as these temporary uses as living or storage space do not turn into projects for the progressive development of the premises. The rightful owner would inevitably see this as an attempt to permanently occupy a space that might be beyond his or her control, whereas as long as squatters occupy a dilapidated space without seeking to improve its comfort, the owner generally sees no major inconvenience. This attitude applies to monuments and/or memorial spaces under the responsibility of municipal or state authorities as well as to privately owned buildings. Houénoudé (2013a, b) shows that the coexistence of two (sometimes contradictory) land tenure systems was at the root of some of the land contradictions observed in the management of the city.

We can conclude that the will of a public or private institution or even an association, in all cases authorized, can transform a place into a space with an economic purpose, the income generated being shared between the family owning the property and the body managing it, with a fee being passed on to the state or the municipality. Otherwise, the property is simply occupied, with tacit authorization or by verbal agreement, either by a person responsible for guarding it and whose role is to prevent any "wild" squatting, or by the poorest members of a family, the majority of whom live outside or in another town in the country who, lacking the means, are offered accommodation in a run-down building, sometimes falling into ruin.

In any case, the family that owns the building remains present, discouraging any further appropriation, which explains the fact that, despite the fact that artists' collectives are quite numerous, they are unable to invest in these buildings to offer them another (cultural) destination, unless they are members of one of their families.

For example, many houses have a temple inside or in the courtyard (or backyard) dedicated to ancestors and deities venerated by the family. Also, regular cults take place in these temples. The immaterial load here is much more important than the building itself, which explains the fact that the cults still take place even though the main building is severely degraded. The existence of these temples is one of the reasons that induce artists not to occupy spaces considered sacred and places of worship.

6.7 Conclusion

The "aesthetics of the rest" brushed by Crespi (2018) poses an important challenge for the valorization of innumerable urban spaces in sub-Saharan Africa with unfinished landscapes, abandoned, derelict buildings and unbuilt spaces, and finally abandoned building sites. For Amselle (2005), the term wasteland is understood as an alternative place, an intermediate space where new artistic forms are produced. For him, it is from this paradox that wasteland derives its vitality from ruins. The theory of the unfinished then becomes, in our opinion, a key to the fundamental interpretation of the African city, to read innumerable contexts with imprecise and changing contours, which take on new meanings as soon as an individual or a group becomes interested in them, uses them and offers them a new, albeit ephemeral, character. These spaces characterized by temporary uses prove to be flexible, use the elements present, which are project materials, and are recomposed and readjusted with a different design.

In the West African landscapes, it is easy to see that the outside space is, in general, very poorly designed. This observation led AbdouMaliq Simone (2004) to speak of "people as infrastructure".

The place given to housing (at home) in Africa in general, and in West Africa in particular, relegates open spaces to the sidelines. However, this perception is relatively recent and does not apply to pre-colonial Africa, where open spaces had a certain importance in spatial distribution. This is, moreover, what the collective of authors of the Guide of good practices for the development of public spaces (2012) shows when they approach community squares (public and private-vodun) as structuring spaces in the old core of the city of Porto-Novo.

Today, the frequent absence of proposals from the State for grouped rehabilitation projects, including a set of spaces and buildings for articulated functions, which are neither contradictory nor redundant, results in and as a consequence of punctual and fragmented interventions which neither achieve the status of a city project nor create real urban dynamics which can regenerate neighborhoods or parts of the city.

Events, because of their ephemeral qualities, although participating in the cultural life of cities, do not generate major dynamics on a larger scale and over time.

References

Adegbidi V, Azokry-Degnon M, Genard J-L, Houenoude D, Keita B, Lemaire J, Mehou I, Noudaikpon G, Robert Y, Tassi S (2012) *Guide de bonnes pratiques pour l'aménagement des espaces publics. Des espaces publics pour le mieux-être des communautés urbaines*, Les publications de l'EPA, Porto-Novo: École du Patrimoine Africain/Faculté d'Architecture La Cambre-Horta de l'Université Libre de Bruxelles, 148 p

Amselle J-L (2005) L'art de la friche. Essai sur l'art africain contemporain. Paris, Flammarion, 213 p

Coralli M, Houenoude D (2013) La politique de patrimonialisation à l'occidentale et ses conséquences sur un territoire africain: Le cas de Porto-Novo au Bénin. Espaces et Sociétés. Les aléas de la patrimonialisation urbaine, Paris, ERES, pp 85–101

Crespi L (2018) Manifesto del design del non-finito. Postmedia books, Milano, p 60

Dossou S, Sogan R (2007) Projet Maisons d'hôtes et valorisation du patrimoine culturel à Ouidah – Bénin. Rapport de mission, Cotonou, p 47

Houenoude D (2013a) Questions patrimoniales et droit foncier au Bénin : le cas de Porto-Novo. In: Raynaud MM, Diop D, Simonneau C (ed) Repenser les moyens d'une sécurisation foncière urbaine: le cas de l'Afrique francophone. Montréal, Service d'Impression de l'Université de Montréal, pp 95–108

Houenoude D (2013b) La protection du patrimoine à Porto-Novo: l'action de la Maison du patrimoine et du tourisme. In: Ch. Mengin et A. Godonou (eds) Le patrimoine de Porto-Novo: réhabilitation, développement, perspectives touristiques. Paris/Porto-Novo, Publications de la Sorbonne/ École du Patrimoine Africain, pp 299–314

Ministere du Cadre de vie et du Developpement Durable (MCVDD) (2018a) Projet d'aménagement urbain et d'appui à la décentralisation (PAURAD), (accord de financement n°5274-bj du 09 octobre 2013). Mission d'études architecturale et technique, de contrôle et surveillance des travaux de réhabilitation des trois (03) maisons afro-brésiliennes dans la ville de Porto-Novo (Lot mo1706-paurad/mab_pn), Rapport d'études d'avant-projet architectural détaillé, Maison d'Oliveira, 67 p

Ministere du Cadre de vie et du Developpement Durable (MCVDD) (2018b) Projet d'aménagement urbain et d'appui à la décentralisation (PAURAD), (Accord de financement n°5274-bj du 09 octobre 2013). Mission d'études architecturale et technique, de contrôle et surveillance des travaux de réhabilitation des trois (03) maisons afro-brésiliennes dans la ville de Porto-Novo (Lot mo1706-paurad/mab_pn). Rapport d'études d'avant-projet architectural détaillé, Maison Koukoui, 87 p

Ministere du Cadre de vie et du Developpement Durable (MCVDD) (2018c) Projet d'aménagement urbain et d'appui à la décentralisation (PAURAD), (accord de financement n°5274-bj du 09 octobre 2013). Mission d'études architecturale et technique, de contrôle et surveillance des travaux de réhabilitation de trois (03) maisons afro-brésiliennes dans la ville de Porto-Novo (Lot mo1706-paurad/mab_pn). Rapport d'études d'avant-projet architectural détaillé, Maison Reynette, 85 p

Paradou P (2011) L'artiste Edwige Aplogan drape le Bénin. Interview RFI à l'artiste du 09/03/2011

Simone A (2004) People as infrastructure: Intersecting fragments in Johannesburg. Public Cult 16(3):407–429. https://doi.org/10.1215/08992363-16-3-407

Chapter 7
Theaster Gates, Freedom, and the Long View

Charles A. Matz

> *I think I'm a full-time artist, a full-time urban planner, and a full-time preacher with an aspiration of no longer needing any of those titles. Rather, I'm trying to do what for some seems a very messy work or a complicated work.*
> —Theaster Gates

Abstract Creative Systems Innovation lives and breathes within the variegated geographies of the City of Chicago's South Side, despite the manifest failures of state-driven governance. And with vibrance, it pulses through the metabolic strains of the art-based eutopia of Theaster Gates' built environment projects: Dorchester Industries, Stony Island Arts Bank, Black Cinema House, Dorchester Art + Housing Collaborative, Black Artists Retreat, Archive House, Listening House, The Rebuild Foundation. Freedom and the long view, an open call in four parts to contribute in substantive ways.

Keywords Theaster gates · Contemporary artist · Creative systems innovation · Dorchester industries · The Rebuild Foundation · Urban planning · Chicago · African-American art

7.1 Part I: The Gates System of Creative Innovation Practice. An Open Call

The interdependent, kaleidoscopic works of urban reinvention promise freedom via explicit tools of systems innovation practice. Theaster Gates makes new, by creating integrated design platforms through art, unlocking deliverance to his community—and with it—a scalable, exportable, production bible for the resurrection of unfinished

Charles A. Matz (✉)
The School of Architecture and Design, The New York Institute of Technology, 1855 Broadway, New York, NY 10023, USA
e-mail: matz@charlesmatz.com

L. Crespi (ed.), *Design of the Unfinished*, The Urban Book Series,
https://doi.org/10.1007/978-3-030-73457-2_7

spaces. Dorchester Industries, Stony Island Arts Bank, Black Cinema House, Dorchester Art + Housing Collaborative, Black Artists Retreat, Archive House, Listening House, and The Rebuild Foundation.

We call on intrepid, multidisciplinary professionals to engage, discover, document, and articulate the results of Gates' achievements. And further, to support implementation of the system across other communities with the aim of tracking a repeatable, recognized innovation-based concept of equitable contemporary art practice. His portfolio of work is an invitation to practitioners of design, systems thinking, economic theory, history, strategy, and innovation practices to explore a new paradigm of organization. One that is singular to the history of art, the American landscape, and located squarely within the context of the vibrant cultural traditions of the founding and extant citizens of the United States of America of African descent, The Community.

Systems Innovation in the sphere of management, technology, and economic theory can be viewed through the lens of dynamic transformation. And in the context of the arts-based practice of Theaster Gates, we can extend this perspective further to include forgathering, synthesis, and redeployment of forsaken ready-parts within the framework of fragmented cultural memory. Within his premise of continuity of inhabitation, institutional foundation-making, platform synthesis, and systemic complexity, lies the promise of a new convention of space-making. The Gates practice has proven to garner power and assignation, within The Community and within planning circles. His projects provide tactical methodologies for scalable systems of civic engagement through culture clustering, and shared infrastructure creation; all woven from the strands of a unique brand of systems innovation practice.

Thaester Gates, who built his early career within the arts, creates, owns, and operates a unique portfolio of projects. As an evolving collection of growing, interrelated

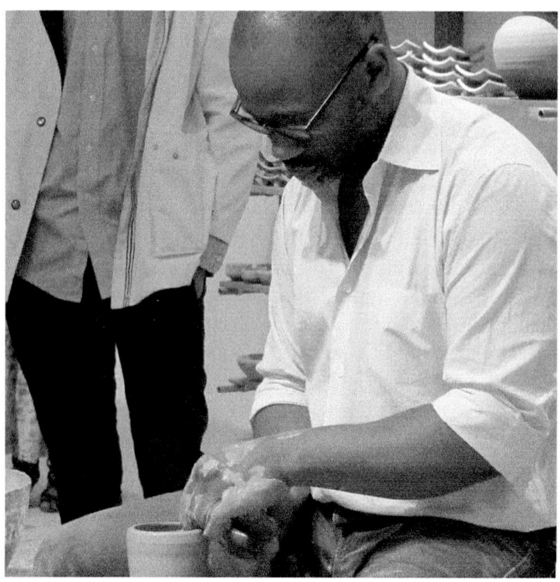

Fig. 1 Theaster Gates, *Soul Manufacturing Corporation*. December 6, 2012, in the Public domain. Photo source: Locust Projects

work, they define a singular program of what can be termed creative systems innovation, and within the context of this subject, the built-environment creative systems innovation of the unfinished.

The significance of Gates' work lies in its relation to the institutionalized American system of governance vis-a-vis The Community. The work stands as a response to the absence of a redress strategy for American citizens of African descent; which is made manifest through the configuration of extant, legacy power structures, the inertia of inopportunity, and well-documented inaccessibility to capital. It also stands in opposition to the absence of participation by the other dominant segments of society—Americans of European descent and assimilated immigrants who regularly access the economic and political frameworks of the United States. The strong interdependence and interrelationship between the systems and subsystems of governance require a plurality of participatory factions for national success. The community served and the communities adjacent to Gates' projects require interface and action-based dialogues toward solutions; by definition, a systems thinking response spurred by complex thinking and communal exigency. Gates, following the historic precedent of self-reliance, has re-formed a centuries-old tactical plan of action, designing systems from discarded parts, disenfranchised coalitions, and within interstitial spaces. Urgency, in this case, is recast with methodologies of contemporary arts practice. And more specifically within the canon of African-American cultural tradition of creative arts production. The very basis for the innovations of culture The Community has offered the world, in myriad forms, and to which the United States has laid claim to.

7.2 Part II: A Proposal for a Further Look at Creative Systems Innovation

The concerned, multidisciplinary design professional can begin their journey of participation by entering the ground floor of the discussion. With narratives of recognized systems thinking in the mold of R. Ackoff and scholars of his generation, the theorists and practitioners led the charge by championing the expansion of systems thinking in resolving complex challenges. Their focus was on the synthesis of elements, relational dynamics between system sets, and the environments surrounding them. System innovation, in scholarship and business practice, flourished on the principles of a new generation of practical viewpoints. Soft and hard systems approaches, Holism and narrative-style solutions to civic problems. Irrespective of the success of these schools of thought in complex problem-solving, the basic evolutionary track emanated from the halls of dominant, Euro-centric institutions. The realities of these centers of study may hold little relevance to the lives of citizens of the United States of African descent. And even less so to the transmitted, legacy-focused African traditions of philosophy, religious practice, socio-political governance, scientific experimentation, and artistic production methodologies; the gamut of cultural memory held within The Community.

Gates inhabits the domain of contemporary innovation practice. More precisely, within the confines of problems with well-defined challenges and concatenated, elusive solutions. Where competencies and resources are emergent, tracking against the perpendicular vector of clear, necessary outcomes. The emergent condition, by circumstance, is a product of multiple, deliberate systemic loops acting in concert to reinforce ongoing fracture. When standard approaches have failed The Community in countering ever-receding equity, radical recombination becomes the normalizing force. With creative art production practice, in the African-American tradition, leading the experimental edge.

7.3 Part III: The Elements Made Manifest

The intrepid built environment, design professional can speculate on the employment of contemporary art techniques in creative systems innovation for the unfinished by exploring the territory further. What can the physical strands of Gates' creative innovation practice do for other, ill-defined, intestertial, spaces? As a third stage of engagement and support, the participant benefits from instruction gleaned from the artist's two online portals. These hold the implicit organizational designs to the expounded methodologies of engagement. They are derived from Gates' own communications materials; the portfolio, approach, program, and function. Theaster Gates artist portal, and the Rebuild Foundation's listed projects:

The Black Monks

… The Black Monks often function as "amateur historians, senior docents, and non-sponsored bootleg preachers" expounding the word of art alongside the word of god. Through this gospel soul chant reverberation, Gates paints another picture of the potentialities within culturally specific but broadly received artistic practices …

Dorchester Industries

Dorchester Industries is a manufacturing platform that creates furniture, objects, and spaces using exceptional but often overlooked materials sourced throughout the City of Chicago. This project has a dual mission: to create beautiful things well and to train our employees to pursue careers in the building trades and creative industries. The objects produced through the collaborative creative process are sold under the Dorchester Industries brand, with proceeds supporting the mission to promote culture-based, artist-led, neighborhood-driven community revitalization.

Black Artists Retreat (B.A.R.)

"The beauty of the proposal of retreating or the creation of the Black Artist Retreat (BAR) is that, at once, it acknowledges both the importance of people and the importance of land. It's a way of recognizing that artists have the ability to unapologetically return to each other."

Stony Island Arts Bank

… the restored Stony Island Arts Bank provides the South Side of Chicago with 17,000 square feet of space for innovation in contemporary art and archival practice. Archival collections housed at the Arts Bank include

Glass Lantern Slides

A collection of more than 60,000 glass lantern slides acquired from the Department of Art History at the University of Chicago….

Johnson Publishing Company

A collection of books and periodicals donated by the Johnson Publishing Company (JPC), publisher of Ebony and Jet magazines; JPC publications from the 1940s to the present day; and the inhouse library once used by JPC editors and writers.

Frankie Knuckles Collection

The personal vinyl collection of the godfather of house music. Knuckles was one of the first DJs to perfect the art of mixing records and was known for overlaying electronic drum machine rhythms onto disco tracks.

Edward J. Williams Collection

A collection of approximately 4,000 objects of "negrobilia" that make use of stereotypical images of black people. … "Black Autonomy alone is too Radical for the current America."

Black Cinema House

Black Cinema House hosts screenings and discussions of films by and about Black people and the issues shaping our lives. BCH also offers community video classes for youth and adults, encouraging our neighbors to explore their creativity, tell their own stories, and develop the skills to shape their own media images.

The Dorchester Art + Housing Collaborative

The Dorchester Art + Housing Collaborative (DA + HC) is a rehabilitated public housing project, a block of 32 townhomes that provides housing for artists and community members with the intent of fostering dialogue and collaboration between both groups. The DA + HC is mixed-income housing and features an even distribution of artist, public, affordable rate, and market-rate housing.

The Archive House

The Archive House is located next-door to the Listening House and serves as a gallery, micro-library, community nexus, and one-time administrative hub for Rebuild Foundation.

The Listening House

Originally a local candy store, the Listening House underwent a renovation to accommodate the Dr. Wax records, portions of the Johnson Publishing Library, and remaining stock from the now-closed Prairie Avenue Books, an architecture bookstore formerly located in the South Loop of Chicago.

7.4 Part IV: Integration

It's worth noting that the portfolio of physical interventions and their associated activities integrate across established classifications of systems. The intrepid built-environment design professional can engage the projects with a shared vocabulary of systems thinking. The widely circulated categorization work of Peter Checkland proposes natural, designed, physical (artifact-based) and abstract (purpose-driven), human activity (mission-focused), and transcendental systems. These are all present in the projects. They are also allocated, in proportions of emphasis, to the foundational track of Gates' trajectory, from land governance to monastic practices. The task ahead is for more artists and designers to contribute in concrete ways to the systems work, actively countering tendencies of exploitation. Action can occur through extraordinary resource allocation, access, constructive regulatory action, legal advocacy, and equity defense. To celebrate the body of The Community's legacy, creative entrepreneurship, and transcendent knowledge with deeds.

> There's a way of believing in the world that forces the artist and the creative in a culture to think about materials in a loaded way. That predetermines how they will handle things, and then how those things will live in the world after. ... How do I make meaning and make sense of the amassing of literal tons of raw material that is in the world to be reimagined?—Theaster Gates. Interview by author Tom McDonough, Bomb Magazine, Dec 10th 2004

Gates' contribution to the world of the unfinished is in the radical reassembly of creative artistic practice, systems theory, and innovation. Design practitioners readily benefit from the robust body of knowledge aggregated in the cultural forge of Americans of African descent. The citizens who built the nation's infrastructure with forced labor, generational blood, sweat, and tears, through ingenious incantation. What are global designers of the unfinished willing to render in return to Gates and the Rebuild Foundation toward freedom and the long view?

Chapter 8
From Grain to Gold: The Regeneration of New Delhi's Dhan Mill Compound

An Industrial Grain Storage Facility Transformed From an Insolvent, State-Owned Property to a Creative Hub in the Service of Urban Fashion and Ad Hoc Design Communities

Mitaali Katoch

Abstract R. K. Jain, the owner of a family run business of agricultural grain mills, storage, and distribution in the state of Chattisgarh, India, packed up his local business and shifted to the capital, Delhi. In the same year, c.1960, he invested in a property located in South Delhi, Chattarpur and launched a new paddy field storage facility. The venture soon became a wholesale distribution hub, the Dhan Mill Compound (*dhan* is the colloquial, Hindi word for grain). Stored grains were supplied to retail shops within the city. In the late 1990s, the company licensed warehouse space to several multinational companies for the storage of their own products. The facility, which ran at a loss for more than six years, transitioned into a creative hub, yielding a sustainable, mixed-market business model in the service of fashion and as a centre for urban design communities. This chapter tracks the history of the previously abandoned Dhan Mill Compound area, and the forces affecting its resurgence as a creative hub. The narrative speculates on the isomorphism of such development projects, within the context of leftover spaces, across continents, driven by repatriated commonwealth communities between India and the United Kingdom.

Keywords Urban regeneration · Leftover spaces · Creative use · Isomorphism · New Delhi and London

8.1 The Design of the Unfinished

Many of Delhi's unfinished spaces are put in service of India's aspirant class, and are inhabited through the creation of hybrid, transnational social settings. These interstitial spaces are modelled and deployed to support the values of the wider, Post-Partition, residual commonwealth ethos. The local ones are components of others,

M. Katoch (✉)
Managing Director of Development; Company Associate, Charles Matz, New York, USA
e-mail: katoch@charlesmatz.com

L. Crespi (ed.), *Design of the Unfinished*, The Urban Book Series,
https://doi.org/10.1007/978-3-030-73457-2_8

similar in nature, located in various cities in India. They are created as mechanisms to reinforce and negotiate, social tensions between populations of repatriated, middle and upper-class professionals. They function as transfigured locations of access to ready-mnemonic images of their time abroad. These designs form concise and repeatable tropes, resolving underlying dichotomies of cultural identity within the context of a landscape of sometimes irreconcilable cultural differences. The Dhan Mill is a crystalline emblem of the growing phenomena. The design of the unfinished as active, socio-psychic resolutions.

8.2 The Dhan Mill: A Prologue

The closing celebration of the 2018 India Art Fair was held at the Dhan Mill Compound. Globally connected, contemporary artists and prominent figures of the New Delhi culture sector mingled within the rugged walls of an enigmatic development. Dhan Mill Compound was billed as a must-visit place (Krishnamachari 2018). Hosted by the founding director of advertising agency Motherland, the glamorous ecosystem was made vibrant by the New York's Asia Society-awarded, Kochi Biennale Foundation's (KBF) founders, the recently appointed British Asian Art Fair director, Khoj International Studio's founding member and director, the trustee of KBF, and Manish Arora—international fashion industry's "John Galliano of India" (Alexander 2007)—who was also once the Creative Director of French fashion house, Paco Rabanne.

Dhan Mill Compound is host to retail stores, cafes, co-working studios, studios for dancers, photographers, performance artists and startups. Classic arbitrage is at work in these spaces. The majority of the lessees and retail store owners have been educated in Europe, the UK or the US. Aman Khanna, founder of Claymen and Infonaut, an alumni of the London College of Communication, set up his art practice in 2005. His studio focuses on sociological themes expressed via clay figurines. The works gained notoriety, while on the shelves of The Plant Society in Australia, and Gallery Bensimon in Paris. Gautam Sinha, founder of Nappa Dori, a luxury brand catering to travel goods, expanded the retail brand into a cafe. He began his business in 2010, and his hospitality concept attracted expatriots and well-travelled Indian residents. His products incorporate Indian and Western tastes.

8.3 The Facility, Mill, and Compound Typology

It's worth noting that the compound itself was a familiar typology amongst other agricultural facilities in India. Grain storage facilities in ancient India were created to save stock in times of famine. During the Harappan period, grain was stored as an exchange mechanism, which was then disbursed to pay taxes, or pay workers (Biswas 1991). There were different typologies of storage facilities such as large granaries, storage pits, and storage jars. Private owners stored grains in common practice. Successive Indian rulers maintained the system taking charge to see to it that surplus was available in times of need. As granary storage areas were privately held, they suffered the vagaries of personal management goals. Hoarding became a common practice. Provisions were made to mitigate hoarding through anti-corruption measures. In 1964, the Food Corporation of India was established by the Indian Government. The organisation became responsible for procurement, storage, and distribution of grains through its licensed warehouses (Pillay and Kumar 2018). These structures continue to exist and are viewed by the local farmers as distribution hubs deferring inflation and to ensure that the farmers' produce is procured at the Minimum Support Price (MSP). During c.1960s, R. K. Jain, the ancestor of present owner of Dhan Mill Compound, wrapped up their business of mills in the state of Chattisgarh in India and moved to Delhi. We speculate from materials acquired in an interview, that there is anecdotal evidence that the Compound received its license to operate under the FCI as it continued to serve as a grain storage facility until the late 90s. The agricultural economy around Delhi waned and the activities of these compounds became outdated. Delhi has since shifted into a service-based economy, its future being the rise of the gig economy in India (Kathuria et al. 2017).

8.4 Economic Development: India

General economic development in India can be divided into three distinct epochs (Mukherji 2009). The period between 1947 and 1968, a period of slow growth during which the government policies were in favour of state-controlled, nationalised businesses. In 1964, the rupee was devalued. Select industries including cement and steel were deregulated. During the ascendancy of India's first woman Prime Minister, India saw a period of disruption with two droughts—one in 1964/65 and another in 1965/66. War erupted between India and Pakistan in 1965 during which India became dependent on USPL480 wheat from the United States prior to a second wave of devaluation of the national currency. The second period falls between the years 1969 and 1974. Under the same leadership the government nationalised private-sector assets such as insurances, banks, coal, and wheat amongst others. Small businesses flourished during this time as there were fewer regulatory bottlenecks and availability of credit. Despite the collective efforts, the economy did not pick up and social unrest ensued. In 1975, the country proclaimed a national emergency with arguable authoritarian characteristics. The political party in opposition, the Janata Party, came to power. The third phase, from 1975 to 1990, saw a period of moderate economic growth. In 1980, industrial deregulation favouring private-sector businesses was made policy. In 1991, industrial de-licensing allowed private companies to produce goods with autonomy. After the trade liberalisation laws were introduced in the 90s, there was a shift from an agriculture-driven economy to a service-driven economy. India has struggled with economic disparity between the rich and the poor. And with colonial-era legacies of favouring conglomerates, most individuals who stood in support of the political party in power enjoyed the benefits of affiliation (Mukherji 2009).

8.5 Context of Economic Development: Delhi

The informal workforce in India has boomed, with more than 90% India's workforce classified as informal labour (Kathuria et al. 2017). These include proprietary and partnership enterprises in the non-agricultural sector or agriculture-related activities excluding crop production. The gig economy has affected all aspects of economic growth in India and especially in the capital. If we look at the current landscape of tech-driven activity allowing transnational trade of services; people can work from anywhere in the world. Companies like WeWork and Regus who offered space to this growing demand became models of development. The gig economy has a significant economic impact on the GDP. According to the Online Labour Index (Kässi and Lehdonvirta 2018), the top occupations that India is known for are software development and technology, creative and multimedia production, and sales and marketing support. Many of these workers, businesses, and startups are in Delhi (Mathur 2020).

8.6 Gentrification and Aspirant Strongholds

The resultant physical expressions of unfinished spaces in Delhi suggest that Western-educated Indian nationals who repatriate tend to recolonise these areas in the image of UK memes. They become touchstones and hubs of social interaction in the service and in support of strengthening the networks of middle-class aspirants. Local custom and use is edged out in favour of economic progress and hegemony. Transnational gentrification is direct and graphic. The devices of the programme include cafe culture, international contemporary research and innovation studio like SPACE10, and global design studios. Craft-art-styled businesses. The discussions in academia centre on the aftermath of colonialism. Gentrification in India is made real by a new, unstoppable wave of internationally styled developments. The post-colonial power dynamic still stands but the current attitudes go beyond the well-worn debates of Indians as postcolonial subjects; these lead to an opportunity of the creation of new global pathways where the choice of accepting or denying the paradigm is in individual hands as opposed to a dictated social need.

8.7 Development: The Arts and Culture Scene in Delhi

Modern-day Delhi is strewn with fragments and unfinished interstitial spaces of its former glory. And irrespective of Delhi's colonial-era power base formerly residing in the hands of the British Empire, cultures survived and thrived, and are still present within the fabric of the city. From the Qutb complex to Humayan's Tomb, the very essence of each passing administration is present in its art. And its former institutions live on. Each era (the Tomar Dynasty, Chauhan Dynasty, Turko-Afghan Sultanate of Delhi, Mughal Dynasty, and the British Empire) left its distinct stamp on the city and has refracted a unique set of socio-political strategic goals. And each has left an art programme to go with it. In the modern period, Post-Partition to the present, Indian art history has been recrafted away from Western modalities by nationalist scholars to steer away from comparison with the European canon of art production (Singh 2017). Intellectuals reaffirm the complex nature of Indian cultural plurality. The current, contemporary art landscape, with its accompanying patronage and production, is unique and reflects the various cultural tendencies of the past. Contemporary organisations have sprung up to deploy a new-found, independent voice of the previously appointed stewards of culture. The art community in Delhi actively promotes the linkages to tradition in order to carry forward the work of these early Indian scholars (Singh 2017). And they, as a community of power-brokers, have organised to protect the cultural icons and history, via referencing and demarcating hard identity boundaries. These institutions have risen up, taken residence in unfinished areas in response to a need for continuity. These include a variety of institutional forms. The Sanskriti foundation is a Public Charitable Trust aimed at promoting and preserving traditional Indian arts and culture. The foundation includes a museum, facility for pottery, workshops and an international artist residency programme. Five acres of undeveloped land were set aside to organise a cultural centre and campus. In 1979, The Sanskriti Foundation was established with state-sponsored foundations (Sanskriti Foundation, n.d.). Initial activities included the creation of Sanskriti Awards which identified artists who furthered the development of traditional Indian arts and culture. Several contemporary artists received honours of distinction including artist Bharti Kher who later founded Khoj International Artists Association (Khoj International Artists Association, n.d.). Delhi throughout the 2000s had a number of galleries without a centralised site for international fairs related to the wider art world. Local institutions lacked a singular mandate to exhibit the works of South Asian Contemporary artists for a global audience. The University of Arts, London graduate and Indian resident, Neha Kirpal, founded the India Art Fair in 2008 (Das, 2012). Currently owned by the British events company, Angus Montgomery Arts, the Fair is an annual, pop-up event held on the Delhi exhibition grounds. It has promoted South Asian Contemporary Galleries and provided a platform for International galleries to ply their wares against the backdrop of the cultural diversity of Delhi's historical patrimony.

The relation between art and craft is yet again reflected in another model of usage within Delhi's complex historical landscape of unfinished territories. A mixed-retail model, first of its kind in Delhi, the Hauz Khas Village is home to thirteenth century Delhi Sultanate building. It has a variety of everything related to creative production: design stores, fashion boutiques, art galleries, educational institutions and the first co-working space attached to an existing restaurant in the city. It saw another layer of shift with the rising need for more quiet spaces with co-working facilities. This gap was met by another business, Cercles. A pop-up workplace. Although, the area is fraught with security problems due to its surrounding bars. Lastly, Khoj, serves a similar function within its unfinished shell; a studio-gallery-institution, founded by the contemporary artists Bharti Kher, Subodh Gupta, Manisha Parekh, Anita Dube and Pooja Sood in 1997. Khoj began as a way to support artists who felt isolated and unsupported. And where they could make art, independent of formal academic, cultural institutions and outside of the commercial gallery system. Khoj is an alternative space for contemporary art practice that constantly challenges the established paradigms of Indian art practice.

8.8 Isomorphism Between London and Delhi Infrastructures, the Repatriation Pipeline of Nationals Educated Abroad

Isomorphism refers to the process of homogenization and is widely used in new institutional theory to define increasing convergence between institutions or organisations, (İnce 2017). We can observe isomorphism in the structure and development of the Mill. Its physical impact, presence and socio-political dynamic are the result of the repatriation of nationals educated in the UK. The system of imprint is supported by ready access to educational institutions abroad and exposure to design theories and art pedagogies within those institutions. Education abroad, for aspirant families, equals validation in society, and further in access afforded to the professional markets of goods and services. Mechanisms of isomorphism in the context of Dhan Mill Compound are mimetic and in some cases normative (Ince 2017).

Points of similarity and disparity exist between Delhi and London's varied arts and culture models, urban planning, and design awareness of development. Delhi mimics frameworks of creative spaces in London, for instance, Delhi's Dhan Mill Compound is isomorphic to London's Shoreditch. Similar to Shoreditch, in-between spaces in Delhi are put to creative use by the design, fashion, and art community. Examples of such spaces include Dhan Mill Compound, Hauz Khas Village, and Khoj. These mimic the superficial function of similar spaces in London. London's repopulation of leftover spaces began in the 1960s as a part of a countercultural movement that challenged ownership of land by greening of spaces (Kamvasinou and Milne 2019). It drew a large number of protesters together. These individuals found creative and innovative solutions around clearing of sites where reconstruction seldom occurred. Public artwork was often used as a tool to attract attention to vacant spaces in the city while reacting to the gallery system, both as physical space for the display of art and as a system of power relations (Kamvasinou and Milne 2019). There are also points of differences to the arts and culture model. London's art and culture sphere is more accessible to local individuals as opposed to Delhi's model which is more restrictive and underground (Singh 2017). These concepts are also held in the urban planning of unfinished spaces. The design awareness within communities is divergent. London has a large number of credible, critic-validated design institutions in comparison to Delhi, leading to more exposure and awareness. The reasons are varied. The major factor is the overt political pressure to make the main focus on how to contemporarise traditional crafts. There are only a handful of credible design institutions in Delhi. Few of them have now been accredited by UK-based universities.

8.9 Postscript

Once a granary, now a host to Delhi's art, design, and fashion community, Dhan Mill Compound manifests the leverage and strength of economic power of association and arbitrage. The Mill is in transition, shifting towards higher-end activities. Its fate follows an arc of inevitable economic rise and a unique brand of South Asian gentrification. The forces affecting its recreation as a creative hub are varied. We can imagine that the inception may have been bolstered by the initial change in economic policies of financial incentives, where fewer obstacles for smaller traders and easy credit were made available. Later, liberalisation of trade favoured the private sector and gave businesses such as the Mill autonomy to produce goods and define new services. These activities followed a shift from an agriculture-driven economy to a service-driven economy giving rise to the gig economy at the national level. The key aspect is in the formation of social networks of individuals belonging to the gig economy. All of whom shared the trait of having had an education abroad. As the economy grew, places to suit the demand of a well-heeled demographic became in demand. They included places like the historic fragments of the Hauz Khas Village. After initial, successful attempts to cater to the repatriated nationals, more formats followed. Sanskriti, Khoj, and other organisations met the goal of creative hubs without hospitality features. These were in support of traditional art and craft communities of India—with restrictions and an international residency model. The newer models, including the Compound, now exist as restaurant and bar spaces with a modicum of a creative community element.

This chapter speculates on the role of isomorphism inside development projects, driven by repatriated commonwealth communities. Indian nationals educated in the UK have fuelled the demand for these works. The trend of repurposing was already prevalent within the context of historic, unfinished spaces, and failed economies. Dhan Mill Compound's mission has augmented a critical new direction: the inclusion of luxury-oriented retail and associated functions catering to aspirant communities.

These well-heeled audiences understand the unique face of social validation and are willing to pay for it. They have the financial resources to afford participation offered by its various, current, and fashionable programmes. And they recognise that progress is in the form of a seat at the spectacle of transformation. A change characterised by embracing the image of tradition against the very real power of material gain, global acceptance, and belonging. The Dhan Mill Compound—from grain to gold.

Appendix 8.1

Figures 8.1, 8.2 and 8.3

Fig. 8.1 SPACE10 Delhi was located in a former granary turned culture hub. Photo by Deepshikha Jain. Courtesy of SPACE10

Fig. 8.2 Handpainted signage in Hindi reads SPACE10. Photo by Sheena Dabholkar. Courtesy of SPACE10

Fig. 8.3 A mural by Aquib Wani for The Irregulars Art Fair at our compound, Dhan Mill. Photo by Sheena Dabholkar. Courtesy of SPACE10

References

Alexander H (2007) Stars of India blazing a stylish global trail. The Telegraph. https://www.telegraph.co.uk/news/worldnews/1546733/Stars-of-India-blazing-a-stylish-global-trail.html. Accessed Nov 27 2020

Biswas A (1991) Storage and rationing facilities in ancient India (From earliest times to the early christian era). In: Proceedings of the Indian history congress 52:98–103. http://www.jstor.org/stable/44142571. Accessed Nov 30 2020

Brosius C, Schilbach T (2016) Introduction: "Mind the Gap": Thinking about in-between spaces in Delhi and Shanghai. City Cult Soc 7(4):221–226. https://doi.org/10.1016/j.ccs.2016.03.003

Das M (2012, February 16) From a hobby to big business: The recently concluded India Art fair has brought into focus growing numbers of women who are tapping into the national market. https://www.dailymail.co.uk/indiahome/indianews/article-2101806/From-hobby-big-business-The-recentlyconcluded-India-Art-fair-brought-focus-growing-numbers-women-tapping-national-market.html

Engle E (2019) Cool workspaces: a look inside IKEA's copenhagen-based research lab. https://www.core77.com/posts/84767/Cool-Workspaces-A-Look-Inside-IKEAs-Copenhagen-Based-Research-Lab. Accessed Nov 27 2020

Goyal S (2020) Conversation with Gautam Sinha of Nappa Dori Cocoa and Jasmine—COCOA AND JASMINE Cocoa and Jasmine. https://www.cocoaandjasmine.com/zine/conversations-with-visionaries-gautam-sinha-of-nappa-dori. Accessed Nov 27 2020

Guerrieri PM (2018) Insights into the cities of India, cultural exchanges, and identities. Negotiat Cult 219–226. 10.1093/oso/9780199479580.003.0006

İnce A (2017) Isomorphism in provision of culture: the case of municipalities in Istanbul and their cultural centers. Middle East Topics Argu 39–50. https://doi.org/10.17192/meta.2017.7.6331

Kamvasinou K, Milne S (2019) Surveying the creative use of vacant space in London, c. 1945–95. In Campbell C, Giovine A, Keating J (eds) Empty spaces: perspectives on emptiness in modern history. University of London Press, London, pp 151–178. https://doi.org/10.2307/j.ctvp2n2r8.14

Kässi O, Lehdonvirta V (2018) The Online labour index. http://ilabour.oii.ox.ac.uk/online-labour-index/. Accessed Sep 20 2020

Khanna A (2005) Infonaut. http://www.infonauts.in/about-us.php. Accessed Sep 30 2020

Khanna A (2020) Profile. https://claymen.in/pages/profile. Accessed Sep 30, 2020

Khanna A, Coelho C (2005) Infomen. http://www.infomen.org/. Accessed Aug 28 2020

Krishnamachari B (2018) An art fair to remember. https://www.theweek.in/columns/bose-krishnamachari/2018/02/24/india-art-fair-an-art-fair-to-remember.html. Accessed Nov 27 2020

Kathuria et al (2017) Future of work in a digital era: The potential and challenges for online freelancing and microwork in India. https://icrier.org/pdf/Online_Freelancing%20_ICRIER.pdf. Accessed Oct 26, 2020

Lawson LJ (2005) City bountiful: a century of community gardening in America. Berkeley, pp xv–xvi

Mathur N (2020) Delhi-NCR sees highest number of new startups during January–June. Live Mint. https://www.livemint.com/companies/start-ups/delhi-ncr-sees-highest-number-of-new-startups-during-january-june-11594548546899.html. Accessed Nov 28 2020

Mukherji R (2009) The state, economic growth, and development in India. India Rev 8(1):81–106. https://doi.org/10.1080/14736480802665238

Mukhopadhyay B (2006) Cultural studies and politics in India Today. Theory Cult Soc 23(7–8):279–292. https://doi.org/10.1177/0263276406073230

Pillay DP, Kumar TK (2018) Food security in India: evolution, efforts and problems. Strat Anal 42(6):595–611. https://doi.org/10.1080/09700161.2018.1560916

Roy T (2002) Economic history and modern India: redefining the link. J Econ Perspect 16(3):109–130. http://www.jstor.org/stable/3216953. Accessed Nov 30 2020

Senthamil V (2018) Troupes of Claymen have set up base in New Delhi's Dhan Mill compound. https://www.architecturaldigest.in/content/claymen-aman-khanna-dhan-mill-delhi/#s-cust0. Accessed Sep 30 2020

Singh K (2017) Colonial, international, global: connecting and disconnecting art histories. Art Transl 9(Sup1):34–47. https://doi.org/10.1080/17561310.2015.1058022

Smith A (2012). The art of urban regeneration. Teach Geogr 37(1):24–25. http://www.jstor.org/stable/23755242. Accessed Nov 27 2020

Waldman D, Silk M, Andrews DL (2017) Cloning colonialism: residential development, transnational aspiration, and the complexities of postcolonial India. Geoforum 82:180–188. https://doi.org/10.1016/j.geoforum.2017.04.018

While A (2003) Locating art worlds: London and the making of young British art. Area 35(3):251–263. http://www.jstor.org/stable/20004319. Accessed Nov 27 2020

Part III
Disciplinary Contaminations

Chapter 9
Urban Remnants Become Setting for Events

Davide Crippa

> *The eye does not see things but images of things that mean other things.*
> Calvino (2019, p. 13)

Abstract The so-called "urban remnants" are often considered as sites undergoing identity crisis, however, in such spatial and temporal suspension, they show a great narrative potential. The suggestion of the unfinished, the versatile spaces open to welcome any story or activity together with their charm, fascinate visitors who are caught into a dramatic atmosphere leading to multiple interpretation. These sites hence become the natural stage for different cultural, design and fashion events where the setting is an integral part of the experience. This chapter analyzes how urban remnants turned into the setting for some of the main events during the Milan Design Week over the past 15 years, through a tale of stories telling their strong calling for theatre and how through little setup works they became setting for unusual metropolitan shows/events. *I cosiddetti "avanzi" urbani sono spesso considerati luoghi in crisi d'identità, ma, nella loro condizione di "sospensione" spaziale e temporale, rivelano un indubbio potenziale narrativo. La suggestione del non finito, la versatilità di spazi non connotati funzionalmente e quindi pronti ad accogliere le storie più diverse ed infine la fascinazione di questi ambienti, che sanno coinvolgere i visitatori in un'atmosfera intensa ed aperta a molteplici interpretazioni, ne fanno teatri naturali in grado di accogliere e valorizzare eventi culturali, di design e di moda, dove la scenografia diventa parte integrante dell'esperienza. Il capitolo analizza come gli avanzi urbani siano diventati sede dei più importanti eventi della Milan Design Week di questi ultimi quindici anni, in un racconto fatto di episodi volti a scorgere la loro intrinseca vocazione teatrale, la loro capacità di tradursi con pochi interventi allestitivi in set per inediti spettacoli metropolitani.*

Keywords Regeneration · Event · Milandesignweek · Fuorisalone · Urban remnants · Exhibition

---

D. Crippa (✉)
Dipartimento di Design, Politecnico di Milano, Milano, Italy
e-mail: davide.crippa@polimi.it

L. Crespi (ed.), *Design of the Unfinished*, The Urban Book Series,
https://doi.org/10.1007/978-3-030-73457-2_9

Urban "remnants" are high-potential spaces, which were initially considered as problematic sites to be redefined because of the expansion of the city's development. A vision closely linked to the disciplines of city planning and architecture, which conceived these abandoned spaces as possible sites for new constructions or, at best, for a potential re-functionalization. An attitude resulting from a nineteenth-century vision of the project characterized by a positivist idea that supposes to be able to prefigure everything in an exact way. In the 1900s, the century of great ever faster and continuous transformations the confidence in the discipline wavers, as the limits of determined projects, which are not ready for change, are highlighted.

The incessant change of our time leads spaces to become obsolete and unused even more rapidly and in this scenario the path of temporary use of lightweight structures becomes a real possibility. An almost performative use of spaces which probably has its roots in the counterculture and hence is so well interpreted by the system of creativity, first by the artistic movements of counterculture and then by the system of design and art. Both worlds celebrate these neglected spaces as new temples dedicated to the launch of new company research, experimental exhibitions of objects and revolutions to the rhythm of the music.

This phenomenon, which started in the 90s, has nowadays skyrocketed so much, that specific dedicated internet websites are created in order to map and offer "ex-something" locations for rent, with particular attention to attractive places, characterized by a hidden story to read. This is how the temporary transformations in the neighbourhoods of Tortona, Lambrate and more recently Nolo and Bovisa originated.

9.1 The Spark of Counterculture

The change of perspective when using abandoned spaces goes hand in hand with the change of disciplinary terminology. If we read again architecture and city planning texts from the 70s to the 90s, in regard to the topic of re-functionalization of large urban voids, we repeatedly find the expression "urban redevelopment" then replaced in more recent times by the word "regeneration".

Precisely to this fact is dedicated one of the interventions in the recent book "#regeneration", which underlines "how the two terms overlap each other and that the redevelopment probably has more to do with the exclusively built dimension, whilst regeneration, in its desire to raise the quality of life of the inhabitants, implements a series of actions that are not necessarily only physical" (Crippa 2020, p. 17) and continuing to emphasize how the regeneration can "hold together the many aspects, city planning, architectural, political, economic, social, identity, functional to the development of cities defined smart, sustainable and inclusive" (Crippa 2020, p. 17). An exchange of words which introduces new meanings and opens up to an interdisciplinary scenario where design, from hybrid discipline as it is, plays a prominent role.

When does the idea of using a neglected space exactly change? From who or what is a new idea for the use of these spaces generated? How is the variable "time" appointed as a project element? Probably the spark that turns on the light on a new path comes from the outside discordant voices, from the world of counterculture, the voice of the famous "world upside down" (Bey 1995, p. 12).

In the practice of RAVES, it is possible to find some approaches that guide the use of spaces even today. The Rave (or rave party) or "techno non-stop 24 h, techno party" (Di Biase 2018), as this practice was initially called in Italy, is a "large gathering of young people, during the night, mostly undercover and of transgressive nature, its location is generally disclosed only a few hours before the beginning of the party, to avoid possible interventions by the police. It takes place outdoors or in clubs suitable for welcoming thousands of people, who dance and listen to electronic, house or techno music at an extremely loud volume, and who often use drugs" (https://www.treccani.it/vocabolario/rave-party/). The definition of the Treccani Encyclopaedia describes the practice; however, it does not draw the attention to the initiatory act of temporary appropriation of unused urban spaces, which is the element that captures our attention in the spirit of the research of this essay.

Raves, self-managed (and illegal) music events, were born in the Eighties. In addition to loud and rapid paced music (mostly tekno, techno, acid house, etc.), they present setup environments, performances by jugglers, works of sculpture, handcrafts by craftsmen, plays of light, the access was open and free for everyone; the duration could vary from one night to more than a week. The Rave phenomenon in the practice of excess has a subversive and political activity which is directly linked to the Hippy counterculture and the movement of Travellers, with their large free fairs and approaches the punk, crew and psychedelic rock countercultures. The direct connection with these counterculture movements determines the principles at the core of Raves, which, however, are established in their identity with electronic music and the development of musical instruments and equipment for sound diffusion accessible to many.

The phenomenon reaches Italy at the beginning of the Nineties. After the anti-rave law of 1994 in the United Kingdom, the various "tribes" move towards Southern Europe, arriving in Italy and in Rome, the city was the epicentre of the national movement. The organizations proposed events outside the commercial logic, looking for abandoned spaces among the inner-city most neglected and deserted neighbourhoods, where the weekly ritual of TAZ or "temporary autonomous zone" was born (Bey 1995, p. 12).

Events focused on issues like sociality, solidarity, tolerance and obviously fun, illegal parties "away from the public sphere, in abandoned warehouses and factories in the suburbs... considered as a subversive act and direct action of repossessing sites that once belonged to labour workers, to exploitation, and instead, with the "Rave" setup, wanted to become spaces of freedom, dance, equality and horizontality, where one could lay claim to their full individual freedom" (https://it.wikipedia.org/wiki/Free_party#:~:text=Verso%20la%20fine%20degli%20anni,locale%20da%20parte%20delle%20autorit%C3%A0). An unconventional revolution, outside the box, or perhaps more simply "a permanent rite of insurrection" (Bey 1995, pp. 12-13), according to the precepts of the mindset introduced by Hakim Bey in the cult book "T.A.Z.".

This historical moment is described by one of the protagonists of the events in an interview published in Vice magazine: "There is always a moment in life where one chooses something to believe in… In the early Nineties, a growing number of young Europeans decided to believe in the subversive power of techno and in a certain way to dance it to the bitter end under the effects of drugs. Ravers felt they were the bearers of a new inclusive and anti-capitalist culture, destined to change the fate of the world, to build an alternative future to the sound of accelerated BPMs and empathogens. There were no videos of the nights… nobody cared about who organized it or who was playing. There were consoles-free parties, the DJs were often hidden and, if you brought a camera with you, good chances are that somebody could get pissed. So, free parties were "a virus within the metropolis", where anonymity was a value. At the first parties, not even pushers were there. A good practice was to go there already equipped with drugs to avoid the creation of drug dealing".

Perhaps, the long wave of counterculture is precisely linked to Cyberpunk, where we find Raves as a tool of action, a possible origin for the search of new uses of urban "remnants". I believe that this energic rush of bass and tribal music made the antennas of creativity vibrate, which has always paid attention to all social changes and phenomena of subcultures. A strategy of collecting recommendations becomes founding, for example: in fashion and in entities like Fiorucci, who, from gathering the stimuli of the London subculture (first and then worldwide), draws the ideas of his vision in apparel.

So, the establishment of TAZs, temporary occupation of abandoned areas in large cities:

- With the creation of their momentary autonomy
- With the attack on forms, on the value of money and politics through free music
- With the cult of anonymity and the denial of the DJ as a "celebrity"
- With the self-production as mass-concept to establish an alternative micro-economy
- With the experimentation of states of consciousness different from the daily working one
- With the empathic approach towards a stranger
- With the search of equality in diversity (outside the traditional political vision).

represent the primordial seed of a shift in perception, which is metabolized and re-elaborated in other forms, running out of its subversive charge. Further, confirmations of the conceptual debt of runways and events of the FuoriSalone come from time coincidences, which witness the creation of FuoriSalone precisely with the arrival of the big Rave gatherings in Italy too (after the glorious start of the 80s in the United States and in Northern Europe).

9.2 The Milan Case

The disruptive power of the temporary event in former factories and in the neglected sites of the metropolis interlaces itself with an intuition from Gilda Bojardi, who in 1991 forged the caption "FuoriSalone" and launched the first guide to events (attached to the "Interni" magazine) to describe all the events that took place in the Milanese city outside of Fiera Milano. A crossroad of themes and opportunities, which then creates the successful formula of the current Milan Design Weeks, which experiment an opening to the general public, a proliferation of events and an increase in exhibit venues, in search of the most particular place to associate to the most experimental or exclusive event.

A need to stay "outside" dictated, in the Eighties, by experimental research of the Alchimia and Memphis studio and that ten years later found a more institutionalized form which starts from the showroom network of the big design companies in town, and then constitutes the districts. The studio focuses on the phenomenon starting from the Tortona district and from the ex-industrial spaces of Superstudio has its starting point, to then evolve and explode in the city. A revolution of use, which from Superstudio for the first time invades the ex Ansaldo (now headquarter of Base Milano) and spreads throughout the neighbourhood in "warehouses that once were full of debris and design objects, but nowadays turned into headquarters of foundations, style offices and in vogue press offices" (Ferrarini 2019).

From the Tortona district, travelling through history again, there is an exhibition of the use of locations that can be compared to the theme of "remnants", as highlighted in the following work carried out in the neighbourhoods of Ventura/Lambrate and Ventura/Centrale, from Alcova to Nolo and from Repubblica del Design in Bovisa and Dergano.

9.3 Tortona District

In the Superstudio warehouses, in an ex-productive context, it is possible to find the right ecosystem to create that contemporary phenomenon that nowadays sees ex-industrial locations among the most coveted spaces for exhibitions, parties, fashion shows and contemporary events of every kind. An expressive potential that is seized by designers, who find in these end-of-life production cathedrals the perfect spots, a genuine aesthetic counterpoint to the most daring experiments, the most inconceivable materials and the most astounding colours. A choice confirmed by fashion that first chooses disused buildings as places to give life to runway shows and then ordains them as headquarter, ultimately recovering the decaying buildings.

In addition to the great events sponsored by Cappellini inside Superstudio, we have the admirable settings of Wallpaper, which can transform multiple times the large deserted volumes of the factory into perfect set designs in order to describe refinement and taste. Locations transformed into magical spaces, just like the interventions carried out by Migliore + Servetto in the former factories between via Solari and via Stendhal: in 2000 with "Wallpaper + Giorgio Armani", in 2001 with "Wallpaper* Urban Addition" and in 2005 "Wallpaper* Express. Milan Fast and fabulous: from Beijing to Shanghai with Bombardier". Setups capable of playing with the decaying majesty of big industrial volumes and emphasizing the sense of temporariness, thanks to an aesthetic of magical isolation, thanks to the use of light and construction material too.

Recently, in the Ex-Ansaldo industrial area, in the part that has not mutated yet, the setup of these volumes was adopted to introduce the relaunch of Richard Ginori ceramics in an installation curated by Paola Navone in 2009. Such installation, in a sharp and didactic way, plays with the harmony of the story: object and method of production presented in a former factory. A simple choice, which can add value and ennoble the sophistication of ceramics, in contrast with the harsh environment of production (Fig. 9.1).

The abundance of spaces "awaiting new use" and the presence of important people such as Gisella Borioli and Giulio Cappellini created the conditions for the transformation, first temporary and then permanent, of the Tortona district, highlighting one of the first phenomena of redevelopment (or perhaps gentrification) connected to the system of creativity.

Fig. 9.1 The great single-subject exhibition of Marcel Wanders in the ex-industrial spaces of via Tortona in 2007 (ph Davide Crippa)

9.4 The Ventura Project Experience

The Tortona district witnesses the birth of the Margriet Vollenberg and Fulvia Ramogida organization, which starts precisely by animating the spaces of the Ex-Ansaldo with a personal exhibition of an emerging Marcel Wanders and then subsequently shifts its attention to various locations of the city; always with the aim of connecting the themes of the emerging and cutting-edge design with a selection of places to be discovered awaiting future uses.

The inclination for ex-industrial spaces as ideal locations for their exhibitions leads the organization to move to the Lambrate area, from which it takes the definitive name of Ventura Project (from the name of the street of the headquarter). The "Ventura Lambrate" project started in 2010 and ended in 2018, creating a trend of incredible success for a district which was mostly unknown to Milanese people, thus triggering that dynamic of discovery of the territory together with the themes of design, making the FuoriSalone formula so special. In eight years, Ventura has strengthened the reputation of the unmissable FuoriSalone event, shaping up to be the most groundbreaking event and that "has always aspired to give visibility to international creatives, both newcomers and prominent people, by launching their projects. A catalyst of ideas and projects, as Maarten Baas defined it among the protagonists of the last edition" (Jonna 2020). A platform between Italy and Holland, which recaptures the pioneering spirit of the first FuoriSalone events (30 years ago) and wants to promote independent designers, mixing them with emerging brands and established international entities (Fig. 9.2).

The format of the Ventura District paved the way for others and was copied several times, but it is in the Lambrate district that a real miracle is performed, transforming a neighbourhood full of warehouses left free from industrial activities into a place of new ideas; a remote place and out of the accepted network but a place not-to-be-missed for its design contents. In the case of Ventura, the appeal is entirely given by the quality of the ideas and by the relationship between the background code (our "remnants") and the object code (the ideas expressed in the form of products and more) leaving almost no room for the setup and favouring the choice of less popular and fascinating places and the best topics to relate to in the form of object selection. A blunt, accurate, sharp way of organizing and interpreting a design event of the Milan Design Week.

The flow of the people moved by the events in this part of town is so significant to make the best results of Rave Parties embarrass and the impact is so strong that, slowly, concurrent actions appear and try to intercept the visitors who have come to the area. Among the most interesting initiatives in this direction, there is the "Padiglione Italia" project, where a selection of Italian designers sets up an old warehouse via Oslavia for several editions, expanding the original outer limit via Ventura, Conte Rosso and Massimiano and kicking off the following expansions having as a consequence a triplication of the area and opening up the majority of ex-productive spaces, from warehouses to factories of all types and sizes. As the borders of the district expanded and exhibition spaces multiplied, the blunt aesthetic

Fig. 9.2 The gigantic former factory at the end of via Ventura opened for the 2013 edition of "Ventura Lambrate" (ph Davide Crippa)

code introduced by the Ventura Project is confronted with less harsher attitudes and so, thus keeping a low budget formula and temporary interventions, the set-ups of Padiglione Italia introduce a stronger setup code in the exhibition; an attitude towards the exhibitory project characterized by a common matrix with the major interventions for Wallpaper of Migliore + Servetto, proposed years earlier in the Tortona district.

In 2018 Ventura Project permanently leaves the area of Lambrate that it had launched and transformed with a relative increase in the values of the spaces, triggering a process similar to the one of the Tortona district.

However, unlike Tortona, which becomes the headquarter of new offices, Lambrate does not find its own definitive and permanent transformation. Currently, the Lambrate area is used for music events and fashion shows, even though it lost its sheen and the regular attendance of the first years.

Ventura organization leaves Lambrate and decides to choose new areas, in renewed consistency with its mission, therefore aiming at forgotten and highly fascinating spaces. Since 2017 it introduced the spaces of Magazzini Raccordati (under the

Fig. 9.3 The first inauguration of Magazzini Raccordati by Ventura Project and the birth of the "Ventura Centrale" project in 2017 (ph Davide Crippa)

Central station), a myriad of square metres forgotten by everyone and re-opened for the first time after many years. So "Ventura Centrale" was born and became the new place to visit. Year after year the organization opens up to new spaces that were previously inaccessible, thus increasing the exhibitory surface area (Fig. 9.3).

The formula of taking new spaces to provide new visions on the metropolis has been reinforced after 2018 with the launch of FutureDome, which, year after year, chooses unknown and unexplored locations to be associated with the innovative contents of the great European design schools. The peak was reached precisely with the inauguration of the former Faculty of Pharmacy, an atypical building that had been in disuse for years and where nature coated the walls with climbing plants and where the different pharmacy laboratories give a special character to the exhibitory contents. With the usual ability of designers to combine places and contents, we find transparency projects, exposed in the chemistry laboratories, reasonings of the chemistry of materials that become a project and thus create that short circuit from content to container, which is eventually the stylistic signature of these exhibitions.

The 2018 exhibition inside the former Faculty of Pharmacy is a manifesto in the form of a space that demonstrates how grasping the spirit of locations, how looking at them, how discovering them under the dust of history and of the unused represents a majestic design tool. A teaching that once again reminds us the potential of these abandoned places, capable of being the best possible scenography for events, if used with the right awareness (Fig. 9.4).

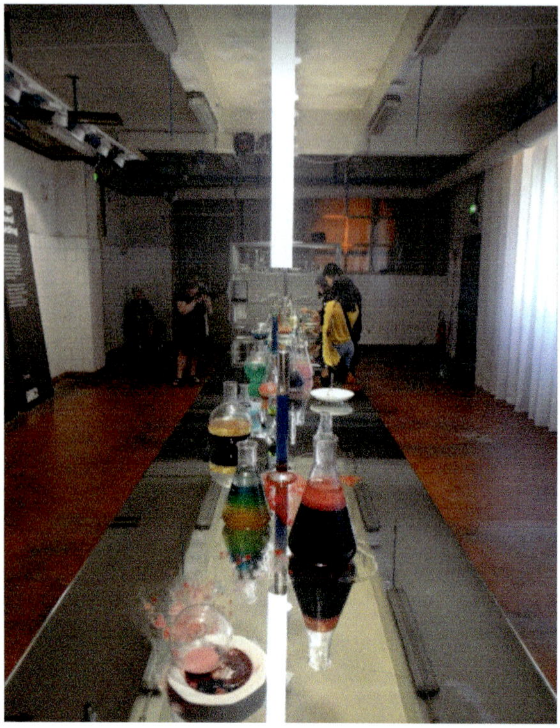

Fig. 9.4 The special opening of the Ex spaces of the Pharmacy faculty with the exhibition "Ventura Future Dome" in the edition of 2018 (ph Paul Pasteris)

Lessons held on being able to choose as a tool for being able to design, lessons that will be confirmed by the previous interventions, like the great "Solo Show" of Martin Baas at the ex-Garage Sanremo via Zecca Vecchia in 2014. Such intervention, a bit exhibition and a bit show, upholds the ability to set up by connecting the places to people and to their own poetics.

9.5 The More Recent Cases of Nolo and Bovisa

To conclude this throwback to history of the temporary uses of abandoned spaces in the design week, we rest on the most recent examples of this propensity that has expanded into the city since the mid-90s. The latest episode in this tradition is the Alcova project, which rolls out the ex-factory of Panettoni Cova in the Nolo district to the city and launches this place in the pantheon of places for exhibiting, walking catwalks and building special events.

The Alcova project, conceived by Joseph Grima and Valentina Ciuffi, since 2018 exhibited for two editions an international selection of designers inside the beguiling spaces of a skeletal factory, where nature entered into symbiosis with the places of production and where the signs of the assembly line interchange with the roots of the trees, that have decided to get their house here. In this space reinvented from desertion, the same strategy described in the Ventura Project is implemented, showing once again the fascination of hidden and inaccessible places of the city and the success that they are able to affirm, if used in the correct way. As in other cases, previously described, the Alcova too will become extremely successful, capturing the attention of this place and, within two years only, it is possible to establish the conditions for a permanent transformation of the area.

A process aimed at promoting voids and abandoned spaces, which can be better developed in the activity carried out by the Repubblica del Design in the suburban area of Bovisa-Dergano-Lancetti. The distinctive element of this new district is the stated willingness to work for the enhancement of the inconclusive places of the neighbourhood, by leveraging on the design and on events as drivers to relaunch the area.

But what are the points of contact and of difference with the cases previously analyzed?

On the one side, the points of difference lie in the awareness of harnessing the temporary project to bring about a permanent transformation and on the other side in using funding announcements for the redevelopment of public places as economic tools instead of renting spaces for the same events.

Apart from these differences in the functioning of the Bovisa-Dergano-Lancetti district, the actions of Repubblica del Design express an attitude towards the "remnants" that traces back to the trends highlighted in the several previous cases.

The activity carried out in the area can be divided into projects like "A letto con il Design" (In bed with Design), an installation proposal that for two editions since 2017 witnessed the transformation of the former vertical factory via Cosenz 44/4 into an odd format of hostel-exhibition for designers. In the hostel project, due to functional as well as aesthetical matters, the theme of the setup distorts the factory layout by re-inventing the use only for the days of the design week (Fig. 9.5).

The attitude is always light, low-cost and reversible, even though it is strongly characterized by its setup dimension. Many of the interventions in the section called "regeneration" (http://repubblicadeldesign.it/regeneration2019/), such as the "Monumento del Design", the "Ultrapiazza" or "Poli Urban Color" fall under the same logic (Fig. 9.6).

Projects like "Though in Milan", "Art 72 room hotel" or the most recent fashion show "HAN KJØBENHAVN—MILAN MEN FW20" and the interventions of the "Design Differente" follow a different path and play with the atmosphere of the places, trying to change as little as possible from what is existing by working on the juxtaposition of the objects in space (Fig. 9.7).

It is possible to evidently confirm, even in these most recent cases, the interest of the event/exhibition system for more disused spaces, so as to connect a trend which started in the 90s in Tortona and keeps on expanding to other areas of the city up to the present day (Fig. 9.8).

Fig. 9.5 Repubblica del Design—“A letto con il Design” in the year 2017, designed by Ghigos for Ideas Bit Factory (www.ideas-bit-factory.it)

Fig. 9.6 Repubblica del Design—“Monumento del Design” in the 2019 Fuorisalone. The installation was designed by Ghigos (www.ghigos.com) with the students of Interior Design of the Polytechnic University of Milan

Fig. 9.7 During men's fashion week, the "HAN KJØBENHAVN" show inside the ex-industrial spaces of the location for events of design differente (www.designdifferente.it)

9.6 Conclusion

Starting from the intuitions of the countercultures, we have outlined a trend, which finds a thirty-year period of alternative and temporary uses of the inconclusive spaces of the metropolises (and in particular in the city of Milan) by the more or less institutionalized design system. An application of spaces, mostly as a setup, capable of hosting exhibitions, performances and runway shows, which progressively spawned more permanent and diverse uses. The imaginative power of these "urban remnants" is so evident that can always influence most of fashion show schedules, and these ones, after 30 years, always choose this kind of places as the perfect frameworks to show off the latest creations of their fashion houses.

This exploration, in the temporary use of urban "remnants" in the Milanese events, allows us to catch sight of two main design attitudes when using these places: the first one is a confrontation and lively discussion among space, setup and exhibited works; the second one is a whispered confrontation, where only the space with the right juxtaposition of works, creates the narrative schedule of the exhibition.

These two different attitudes allow a variety of shades of application, where it is possible to attribute the use that has been made of spaces in these years in the Milanese events and beyond.

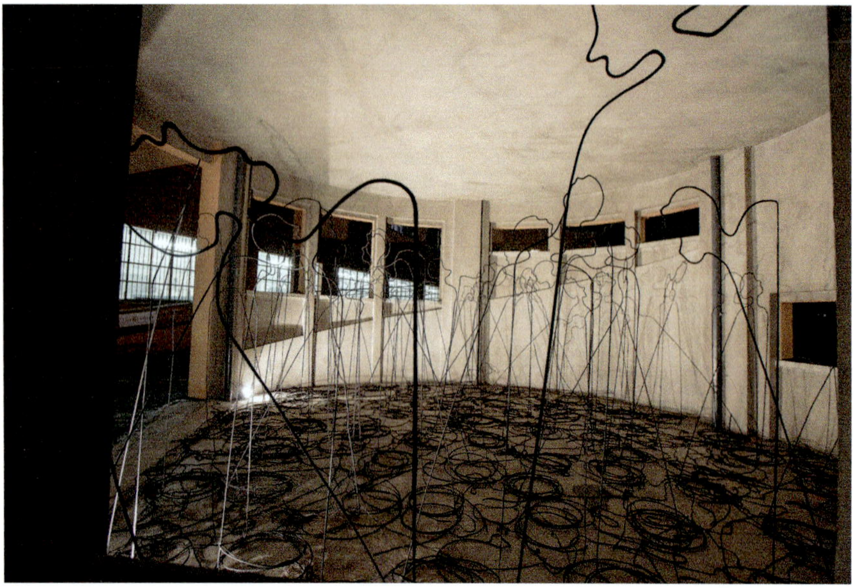

Fig. 9.8 Site-specific installation of the artist Gianfranco Basso in the program curated by Repubblica del Design and Enea on the topic of #energy inside the spaces of Design Differente (www.designdifferente.it)

References

Bey H (1995) T.A.Z. Zone Temporanea Autonome. SkaKe Edizioni, Milano

Calvino I (2019) Le Città, invisibili edn. Oscar moderni Mondadori, Milano

Crippa D (2020) Regeneration. Maggioli Editore, San Carlo di Romagna (RN)

Webliography

Di Biase A (2018) Questo libro racconta la storia dei rave in Italia, Vice. https://www.vice.com/it/article/nepnpk/rave-party-italia-oggi-pablito-drito-libro

Ferrarini P (2019) Fuorisalone: una storia fatta di storie - Dagli anni Settanta a oggi: dentro, fuori e attorno al design, Zero. https://zero.eu/it/news/fuorisalone-una-storia-fatta-di-storie/

http://repubblicadeldesign.it/regeneration2019/ Read on November 10th 2020

https://www.treccani.it/vocabolario/rave-party/ Read on November 10th 2020

https://it.wikipedia.org/wiki/Free_party#:~:text=Verso%20la%20fine%20degli%20anni,locale%20da%20parte%20delle%20autorit%C3%A0 Read on November 10th 2020

Jonna M (2020) Cosa ci ha lasciato Ventura? Un saluto a uno dei progetti più influenti del Fuorisalone, Domusweb. https://www.domusweb.it/it/design/2020/05/18/ventura-project-arrivederci.html

Micciacorta (2016) L'arte dell'anonimato, Tomaso Montanari, La Repubblica. https://www.micciacorta.it/2016/03/21538/

Chapter 10
Beyond Restoration: Reflections for a New Transdisciplinary Paradigm

Claudia Caramel

Abstract In our cities there are many abandoned buildings and architectural details that, although not recognized as official Cultural Heritage, are essential to preserve the identity of places and citizens. As a result of the lack of specific protection measures and social changes, we are losing many of these leftovers and, at the same time, part of our collective memory. While the architectural restoration approach often cannot be adopted to preserve and enhance this heritage, the recent disciplinary debate developed in this area provides many interesting insights to address the issue. In particular, it is necessary to understand what value leftovers can have for citizens and how the use of technologies and migratory phenomena are changing the way people perceive urban spaces. Furthermore, considering the contributions to the understanding of these phenomena developed within different disciplines and the design proposals conceived in the field of architecture, design, and art, this chapter attempts to rethink the theoretical framework within which possible solutions to safeguard urban leftovers can be designed. In particular, the aim is to allow citizens to develop a new awareness of the value of the built environment for our common identity, here understood as a necessary premise for its enhancement.

Keywords Urban leftovers · Restoration · Citizens · Identity · Memory

10.1 Introduction

The changes that have affected the urban landscape since the second half of the twentieth century have left, sometimes unexpectedly, lots of "leftovers" in our cities. Today, they can be recognized in isolated buildings or in details, such as inscriptions reminiscent of an old shop or anachronistic facade decorations, shaped by the desire to improve the aesthetics of the city, now similar to gems set in a foreign body (Fig. 10.1). Silent presences, these urban leftovers seem to be waiting for someone

C. Caramel (✉)
Department of Design, Politecnico di Milano, Via Durando 10, 20158 Milan, Italy
e-mail: claudia.caramel@polimi.it

L. Crespi (ed.), *Design of the Unfinished*, The Urban Book Series,
https://doi.org/10.1007/978-3-030-73457-2_10

Fig. 10.1 The belvedere tower of an old villa is now a leftover in the urban periphery of Gallarate (Italy); Photograph by Davide Niglia

to notice them. However, if we knew how to listen to them, they would be able to tell us a story, which is our common story.

An integral part of the urban fabric and tangible manifestations of our collective identity, these architectural fragments sometimes do not benefit from adequate protection measures. In fact, despite the progressive expansion of the concept of "heritage", that "… has been so over-extended that it now means the opposite of what it originally described" (Nora 2011), many proposals to enhance it still seem driven by economic rather than cultural goals. Today, in fact, the desire to make everything extraordinary leads to the convergence of economic and human resources toward goods with a strong media coverage rather than less conspicuous ones which require long and patient work to be enhanced. As a consequence, the leftovers risk being overwhelmed by noisy architectural and urban interventions that, pursuing "more important" interests, legitimize their demolition and, as a result, the loss of our collective memory.

Often invisible to the eyes of passers-by who move hastily in the urban space, the leftovers suffer from a state of "anesthesia" (Hillman and Truppi 2004) that seems to have numbed our senses. Stunned by continuous visual stimuli and magnetized by the screen of our mobile phone, we notice more easily what surrounds us only when something changes or is missing due, for example, to a demolition. At the same time, the massive use of digital technologies has progressively enlarged our concept of "home" by moving it into a virtual dimension extended to different countries and cultures, but often far from the physical dimension of proximity. The superficial knowledge of our cities and the lack of attention to what surrounds us are the main causes of degradation for urban leftovers and are important aspects to reflect on for their preservation and enhancement.

These remarks can be added to the numerous observations that, since the early 2000s, various scholars have raised in the common need to revitalize the debate on what is the base of the discipline of restoration. In fact, if at the turn of the twentieth century there was a heated debate on theoretical issues, starting from the positions of Viollet-le-Duc and John Ruskin, assumed as opposite poles of the same discourse,[1] in recent decades the discussion is mainly focused on the technical and operational aspects of the discipline. As Paolo Torsello observes, restoration now seems more concerned with *how* to restore than with *why*, forgetting that every technical operation, without a purpose, loses its meaning (Torsello 2005). Furthermore, many terms have been used in recent years to qualify restoration and new professionals have started to work in this area, generating confusion within it. This condition suggested to the scholar the idea of asking the major theorists to give their own definition of "restoration", as an opportunity to reflect on the essence of the discipline. In the same year, Salvador Muñoz Viñas, in his work *Contemporary Theory of Conservation* (2005), examines the orientation of the discipline by proposing "… to deviate the spotlight from the object itself towards the people for whom the object is (and will be) useful" (Muñoz Viñas 2011, 199–200). In fact, as Paolo Martore (2017) points out in the afterword to the Italian translation of this essay, restoration is no longer a question of recognizing the object, but rather of who recognizes himself in that object. Therefore, the focus is not on objects, but on how they are perceived and experienced by people. On this basis, the contemporary ethics proposed by Muñoz Viñas call for *negotiatory* conservation in which both voices of academics and of general public and future users must be heard.

These issues, developed in the field of restoration, raise interesting reflections also for dealing with the complexity of urban leftovers. In fact, only by understanding the value they have for people and trying to imagine the meaning they will have for future citizens, will it be possible to safeguard and properly value this heritage. Moreover, if the lack of protection tools specifically targeted to the leftovers of the building heritage is a deficit, it can open the possibility to experiment innovative solutions that, drawing on different disciplinary areas, offer interesting starting points.

---

[1] According to the thought of Eugène Emmanuel Viollet-le-Duc "Restaurer un édifice, ce n'est pas l'entretenir, le réparer ou le refaire, c'est le rétablir dans un état complet qui peut n'avoir jamais existé à un moment donné" (Viollet-le-Duc 1869). On the contrary, for John Ruskin, restoration "means the most total destruction which a building can suffer: a destruction out of which no remnants can be gathered: a destruction accompanied with false description of the thing destroyed" (Ruskin 1849).

10.2 Urban Leftovers as Tangible Landmarks in Everyday Life

Referring to the questions raised within the restoration field, as a starting point for an enhancement process it is necessary to ask: why restoring urban leftovers? Why keeping these faded traces of a past that we often can't even recognize? In a society more interested in what is extraordinary, what values can these architectural fragments scattered throughout our cities preserve? Why investing in their conservation, in an era in which people live increasingly disconnected from the physical dimension of urban space?

A meaningful response could be read in people's reactions to dramatic events affecting the built heritage. Think, for example, of the areas affected by earthquakes where, when the earth is still shaking, people already call for the reconstruction of collapsed buildings as a way to repair their collective identity (Anzani 2017). Similarly, we could take into consideration the numerous faithful reconstructions of monuments destroyed in the bombings of the Second World War. In this regard, it is enough to mention the striking rebuilding of the historic center of Warsaw, or that of the Frauenkirche in Dresden, rebuilt in the same way through a long and complex intervention which ended only in 2005 with the consecration of the new church (Krull and Zumpe 2005). However, if interventions of this magnitude can be interpreted as the understandable wish to heal physical and psychological wounds, in other situations similar choices seem instead dictated by an excessive cult toward architecture (Hernández Martínez 2010).

Significant in this regard is the recent case of the Cathedral of Notre-Dame in Paris, partially destroyed by a fire that occurred in April 2019. Suddenly, the French were faced with the unthinkable possibility of losing an icon of their culture. Dumbfounded and incredulous, people from all over the world have followed live the collapse of the spire, already rebuilt by Viollet-le-Duc in the second half of the nineteenth century, and become an essential part of the Parisian skyline. Images of ordinary citizens kneeling in prayer near Notre-Dame have traveled around the world making the power of a symbol visible, even in an increasingly technological and virtual society. Equally significant was the debate that followed the event on whether or not to rebuild the spire and, if so, on how to rebuild it. Designers and artists immediately presented their original proposal suggesting solutions in many cases provocative and bizarre. The daring attempt to graft a work of contemporary architecture onto the sacred body of the cathedral, widely evaluated, then surrendered to a reconstruction "as it was where it was", although the structural complexity, already faced by Viollet-le-Duc, does not make it a simple intervention (Bandarin 2020). The decision, confirmed by the authorities, shows once again how the tangible dimension of architecture is necessary for the community to recognize itself even in the twenty-first century.

The strong identity value of the built heritage also emerged during the isolation imposed by the Covid 19 pandemic, when the forced closure of some city landmarks disoriented the citizens. Consider, for example, the importance given by the Italian press to the closing and reopening of the Milan cathedral, an icon of the city and of all its inhabitants. This, like many similar situations around the world, shows that the community still needs to recognize itself in concrete elements, especially in times of crisis, confirming Salvatore Settis' statement that it is when the catastrophe is most evident that we need our fathers most (Settis 2015).

However, interesting "provocations" developed in the field of design and art show how the faithful reconstruction is not the only way to allow citizens perceive physically a ruined building and enjoy its spatiality. In the Italian context, an interesting example is the intervention carried out on the Basilica of Siponto by Edoardo Tresoldi in 2016. Using wire mesh, the material with which the artist usually makes his sculptures, Tresoldi has partially reconstructed the volume of the ancient building, creating a light and monumental structure at the same time. Here art, architecture, and archeology merge in an openly contemporary work that respects both the building and the landscape. The powerful suggestion created by this "poetic" installation shows how also through the emotions aroused in the users it is possible to strengthen a sense of belonging to a common past.

Although the examples briefly mentioned offer different ways of rebuilding, all of them demonstrate how the tangible dimension of the built heritage continues to be essential to feel part of a community. Consequently, if monuments allow maintain a sense of belonging on a national or global level, urban leftovers could have the same relevance in the intimate dimension of the city or neighborhood. They can become the cornerstones of a human-scale daily narration that, paradoxically, the global citizen still needs. In this perspective, the preservation of leftovers becomes the condition to allow citizens to recognize themselves in the proximity, as a fundamental requirement both to ensure the physical and mental well-being of the inhabitants and, indirectly, the care and respect of places (Anzani and Caramel 2020).

10.3 Understanding Citizens to Enhance Urban Leftovers

The second question that must be asked with reference to urban leftovers is for whom is this vulnerable heritage preserved? If an obvious answer is for our children and our children's children, it is more difficult to imagine how the next generations, heirs of a digitalized and globalized world, will be able to relate to the traces of an increasingly distant past. What value will future citizens be able to give to a widespread heritage that they may not even be able to recognize?

In this regard, it is necessary to consider how the digital revolution is changing the use of physical space and Cultural Heritage by the younger generations (Fig. 10.2). Accustomed since childhood to move in virtual play spaces and to perceive the surrounding reality through the filter of their smartphone, will the young people be interested in taking care of urban leftovers? It is easy to assume that people are interested in something when they understand its value, so the question is how to develop in the new generations a greater awareness about the need to safeguard our "common home" (Francis 2015), also understood as the city in which we live. No

Fig. 10.2 Digital technologies are changing the use of urban spaces and Cultural Heritage by the younger generations; Photograph by Davide Niglia, Claudia Caramel and Emilio Lonardo

solution can be excluded a priori but, on the contrary, the positive results achieved in involving digital natives should be carefully examined.

Consider, for example, the success obtained by the video game *Father and Son* designed to promote knowledge of the treasures of the National Archaeological Museum of Naples and the Neapolitan city. Supported by a noteworthy graphic research, the game is based on the story of a young man who goes to Naples to solve the mystery of a letter left to him by his father, an archeologist at the museum. In this way, the player is invited to immerse himself in different historical eras and in many real settings scattered around the city. However, the most interesting aspect of this project is that, in order to continue playing, players must physically go to the museum. Leaving aside the tourist aims and the results in terms of public involvement,[2] this example shows how digital tools do not necessarily remove people from the physical dimension of the city, but on the contrary, they can become a powerful way to bring the knowledge of built environment and its treasures.

Another aspect to consider is how, in recent decades, the perception of cities and Cultural Heritage have been influenced by consumerist tourism, often focused more on the quantity than on the quality of experiences. Last-minute offers and weekend packages invite us to reach unknown destinations and cities of art in a short time

[2]According to experts, 11 months after its launch in April 2017, the game was downloaded about two million times by an international audience and, in total, the equivalent of a thousand years of life was spent by the gamers, whose average age is around 30 years old. Source: https://www.tuomuseo.it/gaming/game-tourism-in-italia-il-caso-di-father-and-son-a-napoli/.

and with little expense. In addition to this, as tourists, we seem more interested in satisfying our narcissism by photographing ourselves near landmarks and to share our photos through social media than in paying attention to what surrounds us. In this way, we have become accustomed to a superficial use of places and to a little attention to the real problems of their protection.

In this regard, emblematic is the case of Venice that, due to excessive tourism, is slowly losing its identity and its inhabitants, forced to leave the houses in the historic center because have become too expensive. Here, even the variety of functions generally present in an urban center has been lost, replaced by hotels and restaurants for people who stay in the city only for a few hours without living there (Settis 2014). Moreover, the particular structure of Venice, built on stilts, also highlights the environmental damage caused by tourism and, in particular, by large cruise ships. These "monsters", immortalized by Gianni Berengo Gardin's shots, have moved tons of water against the fragile scaffolding on which masterpieces such as the Basilica of San Marco are built, only to satisfy the wish of passengers to look at the city from above. However, even if cases like this one show a lack of awareness on keeping the identity of a place alive, it is worth remembering that many other tourist proposals are instead aimed at a slower and more conscious fruition of the territory. Furthermore, it is easy to assume that due to the decrease in economic resources, it will probably be necessary to reduce our travel in the coming years (Latouche 2007). This change, perhaps already forcibly initiated by the spread of Covid 19, could become an opportunity to rediscover proximity and educate the eye to grasp the surrounding beauty and value.

In an attempt to imagine how future citizens will look at urban leftovers, it should not be forgotten that the next generations will be the result of the migratory phenomena we are witnessing today. Different traditions and cultures will be combined with current ones and will perceive the space of the city from new points of view. In this perspective, the built heritage, which has always been a tangible expression of cultural diversity, plays a crucial role in offering opportunities for meeting and dialogue between the parties. Equally important will be enabling citizens to understand the language spoken by the architectural fragments spread throughout the urban fabric and to dialogue with them. In pursuing this objective, the contributions of social sciences should be combined with those of restoration, and different channels of communication will have to be used in an attempt to design truly inclusive forms of heritage enhancement. Making the urban context familiar and welcoming, so that the little-known elements become an opportunity for mutual enrichment, is a difficult objective, but toward which the designer's work should aim.

10.4 A Cross Fertilization to Build Awareness

The attempt to make future citizens aware of the identity value of urban leftovers through their enhancement, raises the need to dedicate some brief reflections also to how to intervene. In fact, the variety of these elements, different in terms of age and characteristics, together with the need to dialogue with a heterogeneous audience, implies a multidisciplinary approach in which the contributions of different fields—psychology, pedagogy, sociology, anthropology, etc.—and the languages of various expressive forms converge. Therefore, the designer has the task of orchestrating these contributions and devising solutions that are both effective and respectful of the built heritage.

For this purpose, many solutions can be borrowed also from art and design. In particular, the artistic experiences developed since the twentieth century offer us a rich repertoire of resources to draw on in an attempt to awaken citizens from the state of anesthesia (Hillman and Truppi 2004) that prevents them from seeing what surrounds us. Consider, for example, the work of Christo and Jeanne-Claude who, in the idea of packaging the monuments of the cities, found a way to make them more visible to citizens. Through their famous interventions, these artists have taught us that in order to reveal something it is often necessary to hide it, suggesting operational methods that can also be replicated for the urban leftovers. Moreover, there are countless solutions which, modifying the normal perception of the urban context only by introducing simple elements such as mirrors, frames, lights, and chairs invite the observer to look at his surroundings with different eyes (Fig. 10.3). In addition, many other solutions based on low-cost and reversible materials can be used to direct users' attention to noteworthy items. This is the case of the yellow footprints that designer Giulio Iacchetti drew under the commemorative plaques scattered throughout the

Fig. 10.3 Changing the perception of the urban context by introducing simple elements can help citizens become more aware of their surroundings; Photograph by Davide Niglia, Claudia Caramel and Emilio Lonardo

city of Milan to emphasize their presence. In fact, although these elements tell an important part of local and national history, they seem invisible to the eyes of passers-by. Therefore, the attempt of this work was to help citizens to recognize and read them, to keep our collective memory alive also in the daily experience of the city and not only on the occasion of national anniversaries. Moreover, this project presents another level of interest because even citizens have been invited to indicate their places of memory by means of a stencil with the outline of the footprints. In this way, from spectators they have become protagonists in the definition of a common memory map in which different points of view overlap in a rich palimpsest.[3]

Active participation of the public is now a consolidated practice in artistic and design experiences and, although moved by different purposes, can offer an opportunity to create or consolidate a sense of belonging to the places and community. In this regard, a particularly significant example for its "political" value was the project *Legarsi alla montagna* conceived in 1981 by the Sardinian artist Maria Lai for the small town of Ulassai. Charged by the mayor to design a war memorial, the artist defiantly claimed the desire to create something for the living, drawing on the traditions and craftsmanship of the place. Taking inspiration from an ancient legend about a little girl who, chasing a celestial ribbon saved herself from the collapse of a cave, the inhabitants of Ulassai were invited to tie the whole village with a long blue ribbon and bring it to the top of the mountain overlooking the town. In addition, Maria Lai asked people to tie the ribbon near the houses of friends so that the relationships between people could be visible (Pioselli 2015). For the first time, the citizens of a small village participated in a relational artistic experience, even more significant if you think about the time and the cultural context in which it was conceived. In fact, the uncooperative attitude of the inhabitants and the hostility of many people have made the artist's work difficult. For this reason, *Legarsi alla montagna* is still a significant example, useful to address the social complexity of our cities where different cultures coexist.

In addition to strengthening the sense of belonging to a human community, experiences like this can help to perceive the physical space of the city as a more familiar environment and, indirectly, foster a spontaneous interest in its protection. Feeling good in a place, both physically and psychologically, is a research topic to which architecture and design offer many answers that can also be drawn for the appreciation of leftovers. In the same direction can also be read the installation *Textile Field* conceived in 2011 by the Bouroullec brothers for the Victoria and Albert Museum in London. In fact, in an attempt to propose a different fruition of the artworks, the designers have placed in Raphael's cartoons room a sort of soft lawn composed of foam rubber textile elements on which people can walk and lie down to admire the masterpieces comfortably. Perhaps, the most interesting aspect of this work is the attempt to put users at ease, not only physically, but above all, psychologically (Crespi

[3]The project *Ai piedi della memoria* was realized in 2009 and presented, through a video, in the exhibition *Oggetti disobbedienti* held at Triennale Design Museum (Milan). On this occasion was also distributed, free of charge, the cardboard mask with two feet and an arrow that allowed citizens to signal their places of memory.

2013). Therefore, a comfortable condition is intended as a prerequisite to understand the value of what you look at. Likewise, identifying areas in the urban space where the citizen can stop and comfortably observe his surroundings can make a decisive contribution to perceiving the beauty and value of leftovers. In fact, making heritage accessible does not only mean allowing the user to reach it physically but, above all, creating the mental conditions to be able to understand and appreciate it.

10.5 Conclusions

Going back to Muñoz Viñas' thought (2005), according to which conservation should be the result of choices shared by all the subjects involved, it appears evident how the attempt to understand the meaning given by citizens to the built environment must constantly guide design choices. On this basis, the designer will be able to develop proposals that respect the citizens' expectations and the buildings themselves, whether they are the result of an "authorized heritage discourse" (Smith 2006) or simple urban leftovers. In fact, listening and having a constant dialogue between the parties seems to be an essential condition in an attempt to preserve the material evidence of our fragile collective identity, today crossed by unprecedented changes. At the same time, the project proposals should offer a context within which citizens can develop a greater awareness to become conscious interlocutors in the negotiation process invoked by Muñoz Viñas.

In this scenario, the designer has the role of mediator and facilitator between the parties and through his creativity can rework strategies and solutions, developed in different disciplinary areas, to create the necessary conditions for a conscious use of leftovers. Therefore, making the urban context a hospitable place (Anzani 2020), putting citizens at ease by helping them to see the surrounding environment, and involving them in enhancement actions become essential prerequisites to foster the development of real awareness. In this perspective, different languages should be used to dialogue with all users, especially with the new generations. The examples briefly illustrated above suggest how even simple temporary solutions, based on the use of low-cost materials, can offer a useful contribution to start a process for the protection and enhancement of urban leftovers. Furthermore, proposals conceived in this way, far from the delicate technical issues that affect the debate in the disciplinary field of restoration, seem to reflect the vulnerability of urban leftovers and, paradoxically, to better respond to the desire for reversibility invoked in the academic context.

If, as highlighted by Paolo Torsello's work (2005), the possibility of enclosing the restoration within a definition appears almost anachronistic in an increasingly "liquid" disciplinary panorama where different knowledge fields contaminate each other (Masiero 2005), the attempt seems even more forced if aimed at defining a single approach to intervene on leftovers. In fact, in addition to contributions from various disciplinary areas and to the ability to orchestrate them, intervening on the leftovers requires to continuously listen to the city, with its changing scenarios where voices, values and cultures are combined in ever different ways. The leftovers are not finished works to be preserved in their uniqueness and to be looked at from afar but, on the contrary, they are the tangible manifestation of changes that affect society itself, which discards and uses what it needs and preserves what it recognizes as significant. As a result, they need a flexible and transdisciplinary approach, in which various techniques, tools, and strategies are used and different languages are spoken. Intervening on leftovers should be understood as an opportunity to give voice to our common past and, at the same time, to leave an open ending for new possible uses and meanings.

References

Anzani A (2017) Bellezza e complessità. In: Anzani A, Guglielmi E (eds) Memoria, bellezza e transdisciplinarità. Maggioli, Santarcangelo di Romagna, Italy, pp 71–79

Anzani A (2020) Identity. Place Identity between preservation and innovation. In Anzani A (ed) Mind and places. A multidisciplinary approach to the design of contemporary city. Springer, Cham, 267–274

Anzani A, Caramel C (2020) Design and restoration: an ecological approach. In: Crespi L (ed) Cultural, theoretical, and innovative approaches to contemporary interior design. IGI Global, Hershey, PA, pp 68–84

Bandarin F (2020) A che punto siamo con Notre-Dame? Il Giornale dell'Arte. https://www.ilgiornaledellarte.com/articoli/a-che-punto-siamo-con-notre-dame-/133816.html. Accessed July 22 2020

Crespi L (2013) Da spazio nasce spazio. L'interior design nella trasformazione degli ambienti contemporanei. Postmedia books, Milano, Italy

Francis (2015) Laudato si'. Encyclical letter of the Holy Father Francis on care for our common home. http://www.vatican.va/content/dam/francesco/pdf/encyclicals/documents/papa-francesco_20150524_enciclica-laudato-si_en.pdf

Hernández Martínez A (2010) La clonazione architettonica. Jaca Book, Milano, Italy

Hillman J, Truppi C (2004) L'anima dei luoghi. Conversazione con Carlo Truppi. Rizzoli, Milano, Italy

Krull D, Zumpe D (2005) Memento Frauenkirche: Dresdens Wahrzeichen als Symbol der Versöhnung/ Dresden's Famous Landmark as a Symbol of Reconciliation. Huss-Medien, Berlin

Latouche S (2007) Petit traité de la décroissance sereine. Mille Et Une Nuits, Paris, France

Martore P (2017) Postfazione. In: Muñoz Viñas S. Teoria contemporanea del restauro. Castelvecchi, Roma, pp 187–189

Masiero R (2005) Nel definire il restauro. In: Torsello BP (ed) Che cos'è il restauro? Nove studiosi a confronto. Marsilio, Venezia, Italy, pp 149–159

Muñoz Viñas S (2011) Contemporary theory of conservation. Routledge, London, New York. (Original work published 2005)

Nora P (2011) Foreword. In: Anheier H, Raj Isar Y (eds) Heritage, memory & identity. Sage, London, p X

Pioselli A (2015) L'arte nello spazio urbano. L'esperienza italiana dal 1968 a oggi. Johan & Levi, Monza, Italy

Ruskin J (1849) The seven lamps of architecture. Smith, Elder and Co, London

Settis S (2014) Se Venezia muore. Einaudi, Torino, Italy

Settis S (2015) Il mondo salverà la bellezza? Responsabilità, anima, cittadinanza. Ponte alle Grazie, Milano, Italy

Smith L (2006) Uses of heritage. Routledge, London, New York

Torsello BP (ed) (2005) Che cos'è il restauro? Nove studiosi a confronto. Marsilio, Venezia, Italy
Viollet-le-Duc EE (1869) Dictionnaire raisonné de l'architecture française du XIe au XVIe siècle (tome 8). A. Morel, Paris, France

Chapter 11
Nature and Anti-nature. The Fractals of the Unfinished

Eugenio Guglielmi

Bottomless wonders spring from simple rules... which are repeated without end.
—Mandelbrot (2010)

Abstract Urban suburbs and many historical centres are characterized by abandoned human spaces that, having lost their original function, now are no longer identifiable. These "no man's lands", which also evoke the metropolis "non-places", however, contain a great use and integration potential that could be developed through the rules of self-similarity which are present in nature and art, as well as in the territory architectures. The Anti-nature crisis, that began around the 1920s and culminated at the end of the 1970s, can find its exit precisely in the theory of the Leftovers, considered no longer as chaos objects, but as real models according to fractal rules. In their still unfinished dimension, they can integrate and better define their surrounding space, making it reconvertible to countless new functions, both social and aesthetic.

Keywords History · Diffusion · Rules · Reconversion

[1] We moved from villages to small towns, where the organic form of nature was still predominant compared to the built one, to the city, then to the metropolis and the megalopolis, where artificial geometric shapes prevailed, basically changing the image of the territory and the perception of the environment itself.

English translation of the chapter by A. Anzani

E. Guglielmi (✉)
Design Campus, Università degli Studi di Firenze, Via Sandro Pertini 93, 50040 Calenzano, Italy
e-mail: guglielmieugenio@gmail.com

L. Crespi (ed.), *Design of the Unfinished*, The Urban Book Series,
https://doi.org/10.1007/978-3-030-73457-2_11

Fig. 11.1 Peripheral occupation (Guglielmi 1981)

11.1 Introduction

Today's reinterpretation of the city unequivocally leads to declare its failure, particularly in the traditional relationship between centre and periphery[1] (Fig. 11.1). Nonetheless, what emerges is an urban fragmentation, almost a composition-re-composition of a jigsaw puzzle that not even the connections, as branches of the traditional infrastructural systems, are now able to make homogeneous, being replaced by "the geometries of technical backbones (meshes, nets, nodes), that identify environmental specificities through increasingly conceptual routes" (Sgalippa 2002a, b) (Fig. 11.2).

The urban development, that has taken place according to a consolidated districts distribution "by areas", has modulated a fabric of micro zones that basically substituted the function of the historical centre, without, however, providing the life quality conditions which were inherent in the ancient nuclei (Fig. 11.3).

The social exchange between individuals, which at the end of the eighteenth century had already disintegrated in European urban agglomerations as a result of the first industrial revolution (1774), remained confined within rigid class hierarchies, converting into marginalization (Ashton 1948).

Fig. 11.2 Urban infrastructure complexity (Guglielmi 1981)

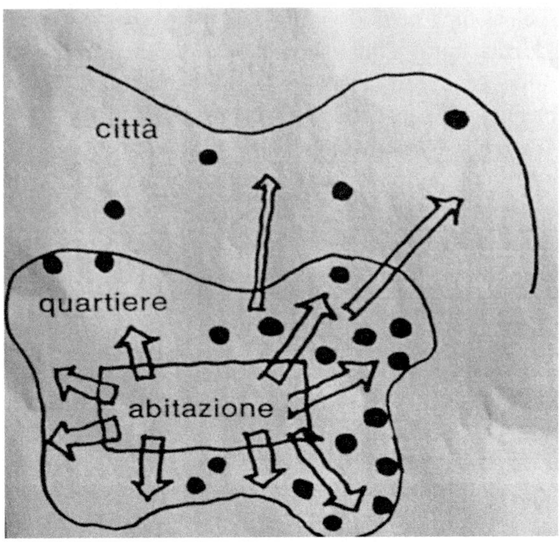

Fig. 11.3 Traditional development by areas of aggregation (Chenut 1968)

Such a legacy had to find solutions in the role of urban planning as a science, aimed at defining the problems related to the development and control of new expansions but, since the 1970s, had already shown its insufficiency with the crisis of environment as a humanized space, in the encounter-clash between nature and man, which Leonardo Benevolo (1923–2017) had defined the "integrated natural space" (Benevolo 1972; Guglielmi 1981, 348–349).

The result of this inexorable process has led our present suburbs and also many historical centres to be characterized by abandoned areas that can no longer be identified, having lost their original function: almost a mocking legacy of the post-industrial "thinned city".[2]

11.2 Expression, Fragment, Document

Paradoxically, these issues were most critically explored in visual art (Klingender 1968). In the nineties of the last century, gathering the results of the neo-capitalist crisis, artists began to deal with social and cultural margins, on the ground of the phenomena of change, difference and inequality, already theorized by the French philosopher Gilles Deleuze (1925–1994) (Deleuze 1968).

A fundamental stage in this analytical process was the Venice Biennale in 1990, titled *Dimensione Futuro* (VVAA 1990), where the operators were called to review their relationship with space without disciplinary distinctions. The result was an obscure and sometimes tragic interpretation of reality that contrasted, for example, with the transgressive optimism of the interventions on the Parisian urban fabric, such as the Louvre Pyramid and the Arche de la Défense.

This almost chthonian vision of the city emerged, in particular, in the exhibition *Ambiente Berlin* where the use of the German word *Ambiente* was interpreted in the meaning of particular atmosphere and of historical and aesthetic setting (Caradente 1990, 59). What is interesting to our reflection is the urban rereading that, on this occasion, did the Dutch sculptor Armando (1929–2018) with the exhibition significantly titled *Remains*, rethinking the "spiritual desolation" of the abandoned areas in tormented cities, unpretentiously rebuilt since 1945 onwards, living alongside the drunkenness of the new at all costs.

Thus, Berlin, the "horrendous city with splendid remains", became a model of how art could fully recover the moors of its Gründerzeit,[3] without necessarily eliminating them, rather reoccupying them in a building key, after the demolition of the Wall in 1989. The recovery of the fragment, of the environmental leftover, its revaluation and

---

[2] A useful synthesis can be found in (Bonicalzi et al. 1976). The extreme consequences of this process of decontextualization have created embarrassing paradoxes such as the construction in China in 2005 of Tianducheng, an exact copy of Paris, including the prefabricated Eiffel Tower 107 m high: a city that has remained uninhabited for years, now used as a destination for newlyweds between reality and fiction and partly occupied by peasants in search of new land.

[3] The period of the German Industrial Revolution.

Fig. 11.4 Emilio Vedova, Fragments, Absurdes Berliner Tagebuch, 1964, detail (VVAA 1990)

aesthetic integration not by chance was dealt with in 1964, by Emilio Vedova (1919–2006) in his studio in Berlin. His extraordinary series of the *Plurimi* addressed the "no man's land" that resulted in the neighbourhoods of the "nefarious Wall", through the use of poor materials with no apparent history[4] (Fig. 11.4).

Again at the 1990 Venice Biennale, even the artists presenting in the Federal Republic of Germany Pavilion, though they belonged to different generations, suffered the power of the landscape metamorphosis and its conservative reinterpretation.

In particular, industrial and railway plants were proposed and investigated, relying solely on the fascination of their productive rationality, as anonymous forms that preserved in the territory could still be able to maintain the human and social spirit that had pervaded them.

Significant in this regard is Bernd Becher's research (1931–2007), which documented these physical presences through a meticulous photographic campaign, interpreting differently their relationship with their places of life between city and production, in the hope of bringing them back to the community.

[4]The jagged and polychrome shapes, the splintering of the painted wood, the intensive energy that emanates from the silhouettes are the spiritual counterpoint that sublimates the faults of men, politicians, of those who see the freedom of peoples as nothing, while instead the carousel goes round and round, inexorably as in Schnitzler's Reigen (Girotondo) (Caradente 1990, 60).

Black and white evocative pictures of ports, traffic, landscapes cluttered of factories bristling with chimneys, of abandoned blast furnaces, references to *The Tentacular Towns* described in the verses by the Belgian symbolist poet Émile Verhaeren (1855–1916).[5]

Là-bas, de nocturnes architectures,

Voici les docks, les ports, les ponts, les phares

Et les gares folles de tintamarres;

Et plus lointains encor des toits d'autres usines

Et des cuves et des forges et des cuisines

Formidables de naphte et de résines

Dont les meutes de feu et de lueurs grandies

Mordent parfois le ciel, à coups d'abois et d'incendies.

Verhaeren (1895)

How much these atmospheres, that seem to overflow from the distant Centre Europe, have fascinated the artists is testified by Alessandro Papetti's work (1958), based on the pictorial representation of disused factories in post-industrial cities, of stationary and abandoned plants, monstrous machines within silent sheds forgotten forever or, perhaps better, bequests ready to become the new temples of our future memories (Pancera 1996) (Fig. 11.5).

English art between 1960 and 1976 provided further proposals on the dematerialization of the urban landscape, useful for the fragments conservation or reinterpretation. This was done in different ways not only involving the architectural project, from Conceptual Art to Land Art, from Poor Art to Post-Object Art, until Language Art and Dematerialised Art, moving from the canonical city spaces to its more indefinite and abandoned areas, indeed considered richer in creative opportunities.

Anthony Caro (1924–2013), Richard Long (1945) or Tim Head (1946) went in this direction, reconverting "real space with art space", redeeming peripheral voids and their intervals for a new social rehabilitation, sometimes only symbolic.[6]

The material collected by the photographic archives of the project *Beni Architettonici e Ambientali della Provincia di Milano (1988–1994)* also documents how in the metropolitan territory of our cities "it is the fragment that wins, both in photographic investigation and evidently also in our personal contemporary life" (Valtorta 1995, 11).

[5]Recognized as "the singer of the city that changes its appearance", Verhaeren had a great influence on Filippo Tommaso Marinetti (1876–1944) and on the concept of the futurist city, the root of the many contradictions of the modern city (De Micheli 1977, 31) (Franz 1952).

[6]See the striking sculpture-sign *Sleepwalk* (1965) made by Caro, *A Hundred Mile Walk* and *Cerne Abbas* by Richard Long (1971–1972), *Installation* in empty abandoned garages by Tim Head (1974) (VV.AA. 1976, 315, 332, 362–363).

Fig. 11.5 Alessandro Papetti, Factory Interior, oil on canvas, 1996, detail (Pancera 1996)

Thus, existential fragmentation becomes now widespread poetics that can no longer be ignored in the aesthetic redesign of the city, starting from urban design.

The *Billboards* carried out in 1964 by Montagna (2009), an environment photographer, are series of images having as their subject the advertising signs found in the deepest Milanese suburbs, such as Bovisa, Baggio, Navigli, Barona, the ring roads that are lost in the fog, "monochrome surfaces that cadence and rhythm these marginal landscapes" (Broggini 2009) (Fig. 11.6).

Just as there are spaces on the borders, there could also be "suburban architects" capable of overturning the apparent process of marginalization of all those expressions that seem to have escaped our rational control.[7]

[7]The term "suburban architects" was ironically coined by Mario Ridolfi (1904–1984) to underline how the opposition to modern culture confined the rationalists' works far from the heart of the city and from a direct confrontation with history (Portoghesi 1974).

Fig. 11.6 Maurizio Montagna, *Billboards* (Montagna 2009)

11.3 The Leftovers Texture as a Value

Ultimately, it is in the continuous quarrel between Nature and Anti-nature where the human perspective is placed, deprived of the correct function of the city and the territory in its basic physical processes, subjected to Anti-nature as a false euphoria of uncontrolled technological development[8] (Fig. 11.7).

Still, in all this, the prophetic echo of the *Somnium Scipionis* of Ciceronian memory seems to reverberate, about the limited man's capacity in the management

[8](Vinca Masini 1978) In our case, Anti-nature means questioning the definition of urban mega-structure, formulated in the mid-sixties of the last century as a derivation of the technological development, in which all the functions of a city or part of it are contained.

Fig. 11.7 Anti-nature. Crossroads of the future city. New York Exhibition, 1939. (E. Guglielmi Archive)

of existential areas, when the earth was still considered as an extension hostile to human settlements.[9]

A process, generated around the twenties of the twentieth century and still evident today, could find its exit precisely starting from the theory of Leftovers (Crespi 2018), considered no longer as chaos objects but as real models. In their unfinished dimension, they can integrate and better define the living space, making it reconvertible to countless new social and aesthetic perspectives, creating a system of socio-cultural networks, different from traditional engineering solutions.

Unfortunately, the Enlightenment concept of order, which still pervades our ideal of a city, continues to condition and limit the choices aimed at its development and transformation.[10]

What does not respond to an economic order based on the value of land and buildings is destined to be abandoned to its fate, thus creating a hierarchy of suburban values that struggle to be recognized as community opportunities.

[9]"Vides habitari in terra raris et angustis in locis, et in ipsis quasi in maculis, ubi habitatur, vastas solitudines interiectas", M.T.Cicerone (106 B.C–43 B.C.), *De re publica* (54 B.C.).

[10]Even the city Utopians, despite their new theories to be opposed to the organization of the historical city, implicitly accepted its old order (Monestiroli 1976, 100–101).

Here, then, is the need for proposals that, together with the exasperated constructive solutions typical of Anti-nature, more and more are able to condition the urban habitat, referring to new anthropological and sociological responses, towards mental territories, even utopian ones, that go beyond the current disciplines of professional order only, to enter ideal multidisciplinary spaces, entrusted to the collective history understood as memory (Le Goff 1992), as a specific characteristic of the Leftovers.

Marco Dezzi Bardeschi (1934–2018) in many of his works has reflected radically on these aspects. In *Plus de Florence S.V.P* presented in 1973, at the Milan Triennale, a "triptych" structure, made with waste materials and human figures reproduced on lateral panels, hosted in its centre an urban plan of Florence as a marginal product of the city that became a representation of itself without time limits (Vinca Masini 1978, 145; Crespi 2017, 197–256). In fact, what is the Leftover if not a proposal between the ephemeral and the space-memory recovery?

11.4 Self-resemblance

The "no-man's-lands", which also evoke the metropolis "non-places", contain a great potential for use and integration through the possible application of the rules of self-similarity. Derived from fractal mathematics, as a recurrence of similar forms on different scales, they are present in Nature and Art as well as in Architecture and in the territory.[11]

Numerous studies have focused on the theme of "non-places" and their relationship with the contemporary city, starting with the term notion as introduced by Augé (1992) (Fig. 11.8). If "lieux", as a positive human presence can refer to the etymology of "logos" in its evident manifestation, "non-lieux" is the negative result of all this, "neither identity, nor relational, nor historical" (Augé 1992).

In the variegated classification by Augé, many of the Leftovers themselves could be included in this interpretation, poised between the loss of their function and their active role in the continuously evolving city.[12]

The peculiarities of these areas, mostly "frayed" compared to the regular urban extension proper to Euclidean geometry,[13] and distributed over the territory apparently without rules as a result of their abandonment, become possible models capable of recreating new physical and social aggregations, starting exactly from their degree of inhomogeneity (Basili 1997).

[11] Self-resemblance is the property of a figure that is similar to a part of it, i.e. having its own shape, characteristic of fractals. Fractal theory concerns non-Euclidean geometry. It was applied by Benoit Mandelbrot (1924–2010) to explain apparently chaotic mathematical behaviour. The term was first used in 1975 and derives from the Latin Fractus, i.e. broken, broken, as well as fractal.

[12] I am referring, in particular, to refugee camps, illegally occupied abandoned houses, bidonvilles, leisure facilities no longer in use, abandoned commercial and industrial areas, etc.

[13] This structure derived from the Hippodamian development, then translated into the Roman thistle and decumanus, continued until our Renaissance through the utopia of the ideal city to arrive in the 1930s to the newly founded cities.

Fig. 11.8 *No man's land*, studies, Indian ink and pastel on cardboard. E. Guglielmi 2017

Therefore, self-resemblance characteristics can be applied to study and control the most advanced and complex urban systems, especially in the identification of hierarchies within those macro-areas that seem to have lost their character of rational unity (Donato and Basili 1997) (Fig. 11.9).

Having overcome the circular and rectilinear city, today urban planners conceive the contemporary urban fabric as amorphous, irregular, with no apparent opposition between "continuity and fragment" (Albini 2016, 25–28).

In 1994, in drawing up the Master Plan for Brescia, Bernardo Secchi defined the great Lombard centre as a "fractal city", precisely because of its territorial discontinuity, in the alternation "of empty and full spaces, fragments of urban facts, each with a specific identity" which had marked "the conclusion of modern Brescia and the beginning of contemporary Brescia" (Benevolo and Bettinelli 1981).

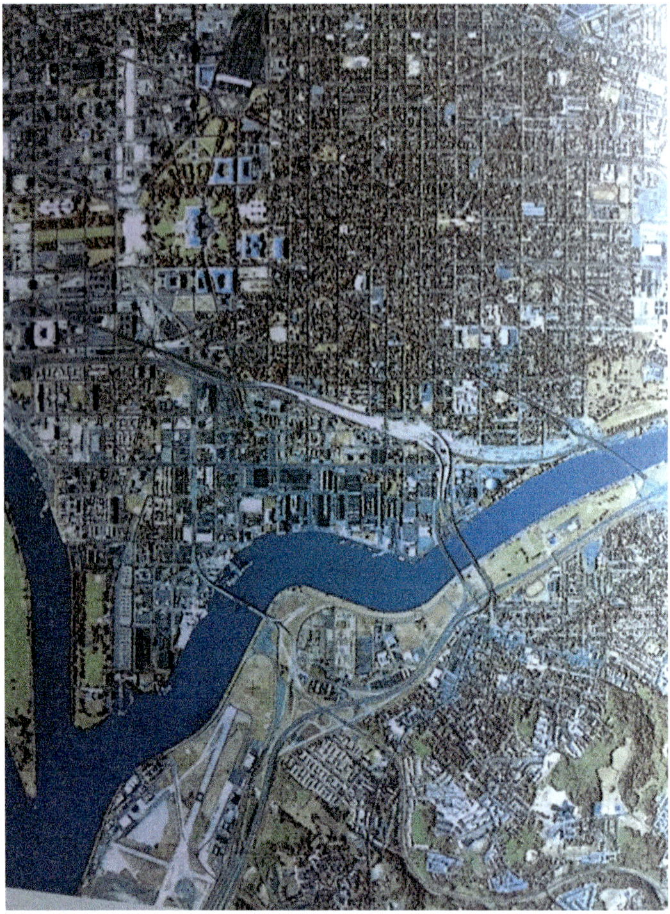

Fig. 11.9 Metropolitan macro area (Guglielmi 1981)

Leonardo Savioli (1917–1982), an architect and painter perhaps too soon forgotten, had already investigated this relationship in the project *Programme of a new urban structure variable over* time (1972–1978). On two non-orthogonal interacting axes, the building meshes are distributed without hierarchical intention scales, no longer conditioned by centripetal development. This results in regular and irregular, linear and broken figures that are repeated in their form and in the same way on different homothetic scales.[14] The difference is cancelled "between urban and architectural space with the use of several languages, even contrasting with each other, of different sizes, scales, *styles*, materials"[15] (Fig. 11.10).

[14]Homothety considers the transformation of a geometric object that from the inside repeats its shape on different scales; by enlarging any part of it, a figure similar to the original is obtained.

[15]Panel on Leonardo Savioli in (Vinca Masini 1978, pp.132–133).

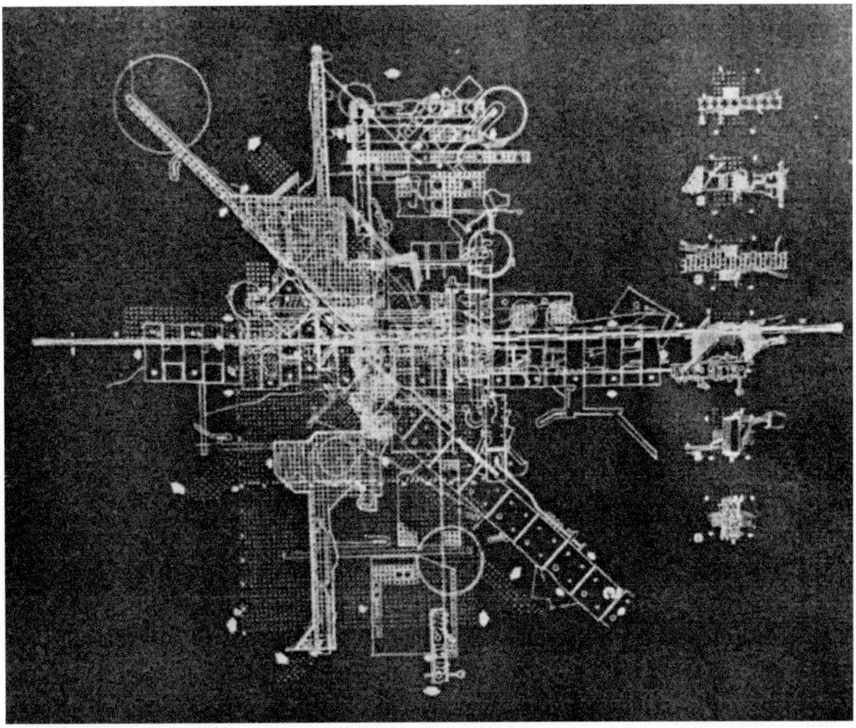

Fig. 11.10 Metropolitan macro area (Vinca Masini 1978)

No longer considered as random or isolated facts, the Leftovers are distributed within a space endowed with homothety, and can thus constitute a starting point to overcome the no longer sustainable binary logic that subdivides the territory into urbanized or non-urbanized.

11.5 Conclusion

In their characteristic of ephemeral "temporariness and reversibility" (Crespi 2019), the Leftovers could thus be treated as "fractals of the unfinished", possible elements of mathematical models that in their dynamic system, through an infinite number of degrees of freedom, evolve over time, overcoming their apparent randomness, according to the cause-effect law typical of the theory of fractal images (Mandelbrot 1983; Peitgen et al. 2004) (Fig. 11.11).

"Objects of a still incomplete dimension", the Leftovers place themselves in their internal and external context as apparently unfinished towards the territory to which they belong.

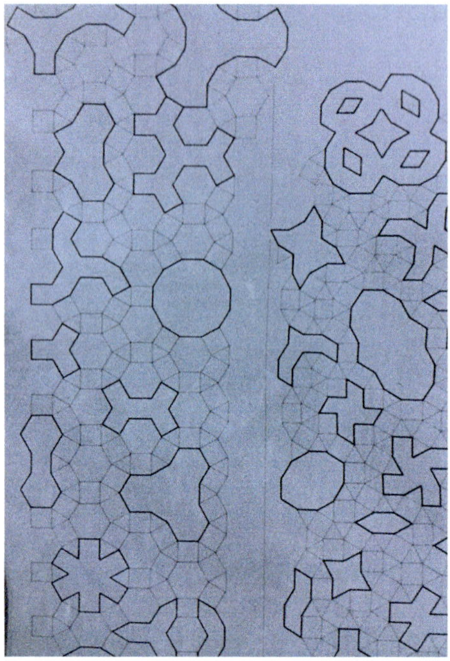

Fig. 11.11 Dynamic reticular system. E. Guglielmi, ISA Monza, 1972

This implies a new design and disciplinary approach on different scales, where demographers, geographers, together with town planners, architects, designers and economists are called to propose their strategy for social mending, through a repeatable and reproducible network according to the place needs, recovering its identity values and in many cases even its forgotten "genius loci" (Crespi 2020) (Figs. 11.12, 11.13 and 11.14).

This is also an opportunity to reflect on the role of new disciplinary skills that, accessing to the "design culture", Luciano Crespi identifies in interior design (Crespi 2018).

Quoting a happy intuition by Gianluca Sgalippa, everything expands and extends, in the environment that seems to bend to the man's will, even the "product becomes a place" (Sgalippa 2020).

Therefore, from the design point of view every intervention, like the plans of territory management, should consider the Leftovers as an active natural resource. A complete census of these fragments should be carried out in their various urban realities, to identify their original values, the reason for their abandonment, their state of decay, material composition, construction techniques, for a reconversion that will find its ideal procedure in the *Manifesto of the design of the unfinished* (Crespi 2018).

This is an innovative way of thinking to those places of our human settlements which are in the shadow, a shadow that Tommaso Campanella (1568–1639) in *The City of the Sun* (Firpo 1997) considered as "a trace of light, sharing light, but not full

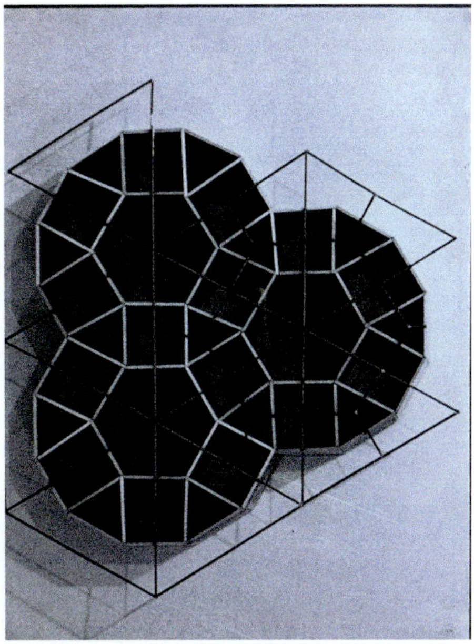

Fig. 11.12 Urban aggregation model referable to a repeatable basic structure. E. Guglielmi, ISA Monza, 1973

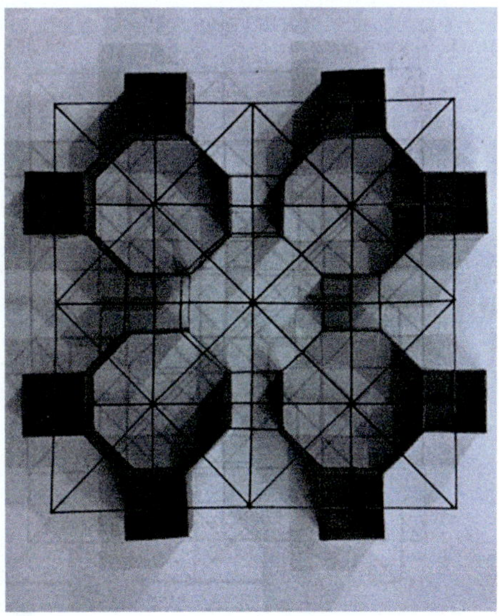

Fig. 11.13 Urban aggregation model referable to a repeatable basic structure. E. Guglielmi, ISA Monza, 1973

light". This is why I would better define the aesthetics of the Leftover theorized by Crespi (2017) as the "poetics of the Leftover" because to give light to the world, we must also be poets (Figs. 11.15, 11.16 and 11.17).

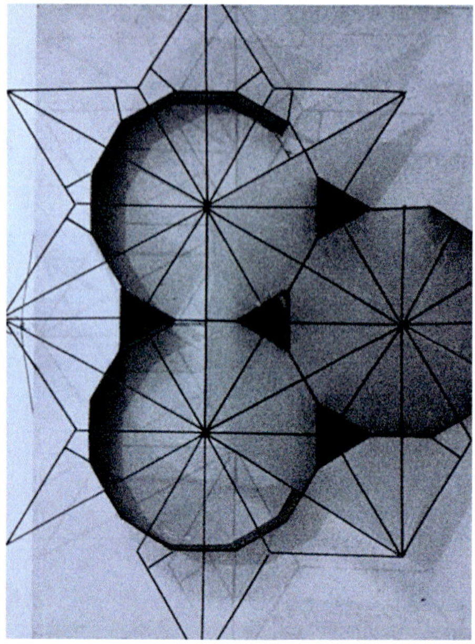

Fig. 11.14 Urban aggregation model referable to a repeatable basic structure. E. Guglielmi, ISA Monza, 1973

Fig. 11.15 Urban leftovers, polystyrene. E. Guglielmi, ISA Monza, 1972

Fig. 11.16 Urban leftovers, mixed media, Indian ink and photographic paper on cardboard. Study for *Hortus conclusus*, E. Guglielmi 2017

Fig. 11.17 Urban leftovers, mixed media, Indian ink and photographic paper on cardboard. Study for *Hortus conclusus*, E. Guglielmi 2017

References

Albini M (2016) *La città contemporanea come architettura mutevole nel tempo*. Collection of lessons from the bachelor's degree course in Environmental Architecture, Politecnico di Milano, Piacenza Territorial Pole, School of Architecture, Urban Planning and Construction Engineering. Lomazzo (Co): Ed. Tecnografica

Ashton TS (1948) The Industrial Revolution 1760–1830. Oxford University Press, Oxford

Augé M (1992) Non-lieux. Éditions du Seuil, Paris

Basili LL (1997) La geometria frattale dell'organizzazione urbana oltre la crisi dell'ordine spontaneo. In: Bertuglia CS, Vaio F (ed) La città come entità altamente complessa, vol 1, pp 205–237. Franco Angeli, Milano

Benevolo L (1972) Le origini dell'urbanistica moderna. Edizioni Laterza, Collana Universale Laterza, Bari, n 91

Benevolo L, Bettinelli R (1981) Brescia Moderna. La formazione e la gestione urbanistica di una città industriale. Grafo Edizioni, Brescia

Bonicalzi R, Del Bo A, Di Leo A, Grassi G, Guazzoni E, Monestiroli A, Prizzon M (1976) I temi dell'architettura della città. Clup, Milano

Broggini V (2009) Rallentate visioni. In: Montagna M, *Billboards*. Catalogue of the Exhibition at the Galleria Manzoni Arti contemporanee e disegno industriale. Magnolia edizioni, Bergamo

Caradente G (1990) Novembre a Berlino. In VV.AA., XLIV Esposizione Internazionale d'Arte La Biennale di Venezia. General catalogue. Fabbri Editori per Edizioni Biennale di Venezia, Milano

Chenut D (1968) Ipotesi per un Habitat contemporaneo. Il Saggiatore, Milano

Crespi L (2017) Estetica dell'Avanzo. In: Anzani A, Guglielmi E (eds) Memoria, bellezza e Transdisciplinarità. Santarcangelo di Romagna, Maggioli

Crespi L (2018) Manifesto del design del non finito. Postmedia Books, Milano

Crespi L. (2019). La rigenerazione degli Avanzi urbani. https://neotopie.wordpress.com

Crespi L (2020) Borders. The design of the unfinished as a new transdisciplinary perspective. In: Anzani A (ed) Mind and places. A multidisciplinary approach to the design of contemporary city. Springer Nature, Cham, 139–157

De Micheli M, Pizzigoni A (1977) Milano 1870–1920: aspetti della cultura positivista e d'avanguardia. Clup, Milano

Deleuze G (1968) Différence et Répétition. Presses universitaires de France, Paris

Donato F, Basili LL (1997) L'ordine nascosto dell'organizzazione urbana. Un'applicazione della geometria frattale e della teoria dei sistemi auto organizzati. In: Bertuglia CS, Vaio F (ed) La città come entità altamente complessa. Franco Angeli, Milano

Firpo L (ed) (1997) La città del sole. Bari, Laterza

Franz H (1952) Emile Verhaeren. Editions Pierre Seghers, Paris. Collection Poètes d'aujourd'hui n° 34

Guglielmi E (1981) In forma d'immagine. Uomo, Arte, Ambiente. Edizioni Atlas, Bergamo

Klingender FD (1968) Kunst und industrielle revolution. Baues Verlag Rainer Baues, Bremen

Le Goff J (1992) History and memory. Columbia University Press, New York

Mandelbrot B (1983) The fractal geometry of nature. W.H. Freeman and company, New York

Mandelbrot B (2010) Fractals and the art of roughness. TED talk https://www.ted.com/talks/benoit_mandelbrot_fractals_and_the_art_of_roughness

Monestiroli A (1976) Il progetto della città industriale. In: Bonicalzi R, Del Bo A, Di Leo A, Grassi G, Guazzoni E, Monestiroli A, Prizzon M (eds) I temi dell'architettura della città. Clup, Milano

Montagna M (2009) Billboards. Catalogue of the Exhibition at the Galleria Manzoni Arti contemporanee e disegno industriale. Magnolia edizioni, Bergamo

Pancera M (1996) Alessandro Papetti. Interni di fabbrica 1993–1996. Catalogue of the Exhibition at the Comune di Lecco, Assessorato alla Cultura, Musei Civici, Galleria Bellinzona. Editrice Stefanoni, Lecco

Peitgen H, Jürgens H, Saupe D (2004) Chaos and fractals—new frontiers of science. Springer, Cham

Portoghesi P (1974) Le inibizioni dell'architettura moderna. Laterza, Bari
Sgalippa G (2002a) Design di soglia. Sistemi di prodotti lungo le interferenze tra sistemi tecnici urbani. In: Sgalippa G (ed) Quando il prodotto diventa luogo. Milano, Franco Angeli, p 133
Sgalippa G (ed) (2002b) Quando il prodotto diventa luogo. Franco Angeli, Milano
Valtorta L (1995) I frammenti del paesaggio e i frammenti della fotografia. In: Sacconi A, Valtorta R (ed) Archivio dello spazio 3. Dall'archivio fotografico del progetto Beni Architettonici e Ambientali della Provincia di Milano. Catalogue of the Exhibition at Palazzo Isimbardi, marzo-aprile. Arti Grafiche Friulane, Fagagna UD
Verhaeren E (1895) Les villes tentaculaires. Edmond Deman, Brussels
Vinca Masini L (ed) (1978) Topologia e Morfogenesi. Utopia e crisi dell'anti-nature. Momenti delle intenzioni architettoniche in Italia. Electa, Milano
VVAA (1976) Arte inglese oggi 1960–1972. Catalogue of the Exhibition at the British Council, Comune di Milano, Palazzo Reale, February/May. Electa Editrice, Milano
VVAA (1990) XLIV Esposizione Internazionale d'Arte La Biennale di Venezia. General catalogue. Fabbri Editori per Edizioni Biennale di Venezia, Milano

Chapter 12
Giving Form to Precariousness

Ermanno Cristini

Abstract The architectural "leftovers" that distinguish the current urban landscape also constitute a focus for artistic research, which is committed to measuring itself with the concepts of temporariness and precariousness, intended as constitutive elements of contemporary culture. Throughout the look for a new paradigm inspired by the ethics of the wanderer, the relationship with architectural "leftovers" can go beyond a static vision of their recovery as containers. In fact, we can look at them thinking about light, mobile and non-definitive interventions. The attempt is to "negotiate" with the space in order to "giving form to precariousness". This way, the artistic interventions would not only rethink the status of work of art, but they would also redefine the identity of the places for cultural, nomadic and other new uses. The chapter analyses these issues by examining two case histories: the exhibitions of Dialogos project (at Assab One, a former printshop in Milan, and at Vitrína Deniska a former shop-window in Olomouch in the Czech Republic) and Sopra-Sotto (a former supermarket in Sesto Calende, in Italy).

Keywords Artwork · Contemporary art · Precariousness · Transitory · Ephemeral · Unstable · Site specific · Time specific · Remains · Leftovers

12.1 Artwork, Site, Situation

Transformations in the city have always represented a challenging terrain for the visual arts as they progressively redefined their very nature. The notions of artwork and place have gradually lost their traditional separate identity, nourishing each other through their mutual expansion and tending towards an inseparable relationship. This reality is even more evident today in the light of the complex paradigms linked to the urban condition, and more generally, to the orientation of skills. Intersection, multiplicity and nomadic indeterminacy have raised new and pressing issues for thought and project, and consequently for the arts, in response to which the specificity

E. Cristini (✉)
Artist , Varese, Italy
e-mail: cristini.reset@libero.it

L. Crespi (ed.), *Design of the Unfinished*, The Urban Book Series,
https://doi.org/10.1007/978-3-030-73457-2_12

of art making can no longer necessarily be defined by rigid categories confined to the object.

Ever since the artwork has projected itself into the surrounding space, it has incorporated display. The "crisis" of the frame has been resolved by going beyond the limit of the painting, to the point that the sacredness of representation has been extended to the place in which the artwork "lives" since it is *put on show*, or, as Giulio Paolini would say, to the place of its "annunciation". "To conclude: the artwork is the body of artworks that are on display. Indeed, the display, in itself and as such, is the artwork" (Paolini 2006).

According to Brian O'Doherty, the problem of the artwork's needing to "breathe", that is to be part of its surroundings, arose in the mid-twentieth century (O'Doherty 2012). The wall evolved from being merely an instrumental element and was gradually incorporated, along with the space of which it was part, into the space of the artwork.

There were earlier known precedents, however, in the first decades of the last century. The laboratory of the historical avant-gardes prepared the ground for the collapse of the traditional categories of painting and sculpture, along with the references to "inside" and "outside" the artwork. In the 1950s and 60s, this gave rise to experiments like *Le vide* by Yves Klein, where the completely empty gallery was the artwork, the form of an immaterial material identified as the "sensibility" contained in the place (Mock 1983). And, in the field of music, to *4′33″* by John Cage, in which the space of the hall becomes music through the "subtraction" of the piano (Cage 2013). Then, just over a decade later, there was Andy Warhol's *Silver Clouds*, in which inflated cushions "floated" in a room of Leo Castelli's gallery, activating it completely from floor to ceiling (Ratcliff 1983). These were followed, in 1968, by Kounellis' horses in Fabio Sargentini's L'Attico gallery in Rome, where the relationship between natural and artificial was expressed, on the basis of the nascent Arte Povera poetics, through the relationship between the horses and the space, the latter no longer being an accessorial element (Barbero and Pola 2010).

Gruppo used the concept of *limit* (Gruppo 2018) to indicate the expedient that delimits the space in which the artwork is enunciated. What occurs in the aforementioned experiments is a "fibrillation" of the limit, to the point that it steadily shifts to include in the semiosis not only the enunciated, but also the space of enunciation. Thus, place in the work as semiosis is inseparable from the relationship through which it is constituted. This, in substance, is the concept of "site-specific".

In the 1960s and early 1970s, this concept became fully established through movements such as Land Art and Minimalism. Carl Andre stated: "Rather than cut into the material, I now use the material as the cut in space" (quoted in Rose 1965). Clearly, this is a strictly physical conception of space, consistent with the Minimalist focus on the phenomenal elementary evidence of the materials. The same thing can largely be said of the Land Art interventions. From Minimalism to Land Art, the trend was marked not only by the shifting of the limit of the object towards the wall, but also by the limit of the gallery itself being crossed, so that the "wall" was ideally extended towards the external space.

While Land Art focused almost exclusively on natural spaces, other experiments broke through the confines of the gallery by considering the urban dimension.

Many of the Situationist, or better still Fluxus, projects adopted this approach, as can be seen from *Galerie Légitime* by Robert Filliou of 1962 (Pissarro and Hunter College Art Gallery 2016). A few years later, on the occasion of the exhibition *Con-temp-l'azione* in 1967, Michelangelo Pistoletto created *Newspaper Sphere*, one of his *Minus Objects*, which was conceived to be rolled along the Turin streets and to disrupt the "normality" of city life in the context of a kind of spontaneous happening (Palazzoli 1967).

Jiri Kovanda's solitary, marginal actions, seminated in a Prague on the other side of the wall in the 1970s, triggered relations more subtly. Standing motionless, arms outstretched to form a cross, in the middle of the street or riding a crowded escalator facing backwards to look into the eyes of the person behind, for example, are minor human shocks designed to restore to the public space the possibility of being a meeting place (Havranek et al. 2006).

We could give many more examples, but what is significant here is that the art object loses any value as such and acquires that of a "device", an activator of involved and involving collective experiences in whose context the site plays anything but an accidental and secondary role.

In fact, we read in the catalogue of *Con-temp-l'azione*: "The action to be performed becomes interaction. It is not things that are important but what is added to them. Things are no longer an end in themselves. They are important for what they produce. And for the relationships they establish" (Palazzoli 1967, op. cit.).

This not only affirms the concept of site-specific, but also extends it to the relational as well as the physical determinants of the site.

Claire Doherty defines this phenomenon as the transition from "site" to "situation" (Doherty 2009) and it is fundamental to the way in which site-specific is understood at the present time, when it is further expanding towards a dimension that also involves the virtual, and is often identified with what is ascribed to Public Art. Miwon Kwon introduces the notions of "site oriented" and "community specific", stressing the purely relational value of the place. "This means that the site is now structured in an (inter)textual instead of spatial way, and its model is not a map but an itinerary" (Kwon 2020).

Cecilia Guida speaks of "Spatial Practices" (Guida 2012), referring to artistic practices that explore a hybrid area, where the notion of public space is transformed through a web-induced, double movement consisting in the dematerialization and expansion of the urban space.

In this context, the artistic intervention straddles online and off-line, consisting of more or less direct excursions into the social, and creating a fringed space. This results in a constant oscillation between rediscovery and loss of specificity within a "vaporized" form, which corresponds to the rhizomatic nature of the new urban landscape.

12.2 "Formed Form" and Relation

The "restless form" of "Spatial Practices" marks the radical emancipation of the concept of site-specific and its related practices from projects that are generally known as urban design. It differs from these in that it goes beyond the separation between work and place, to which the project, in the conventional sense, must offer the possibility of recomposition in a harmonious and stable whole. In urban design, the artwork is intended to *embellish* or complete the site by giving the whole a reassuring homogeneity.

By contrast "Spatial Practices" imply, due to their very nature, a fluidity within which the artwork is not conceived a priori but as an impermanent entity, taking its form from/on the site and from/in its being a situation. The site becomes the material of the artwork and an element inseparable from a "form" whose principal characteristic is the immateriality of relation. Likewise, as evolutions of site-specific, "Spatial Practices" confirm the autonomy of urban art by affirming their implicit critical nature.

In this sense, site-specific means exploring the limits of art making by utilizing "unharmonious" elements of the urban fabric, which correspond to the creation of as many "points of catastrophe", places in which languages, by marking their own crisis, simultaneously determine their regeneration on new foundations. By eschewing the consolatory approach of urban design through its "telluric" nature, site-specific assumes an ethical value, and implicitly, a potentiality to go beyond the project, emancipating from its traditional logic.

In *Relational Aesthetics,* Nicolas Bourriaud notes that art in this key represents a "social interstice", which "creates free and lasting spaces whose rhythm contrasts with those that regulate daily life" (Bourriaud 2010). Moreover, Bourriaud continues: "If we look at contemporary artistic practices, we should speak not of "forms" but "formations": as opposed to an object contained within itself through the intervention of a style and a signature, art today shows that form exists only through interaction" (Bourriaud 2010, op. cit.).

In all due proportion, this vision is deeply rooted in, and in certain respects, tangential to a dynamic concept of form that is indirectly, and certainly not deliberately, grounded in Pareysonian "formativity" (Pareyson 1974), according to which form is a process capable of assuming the means of its formation. "Formed form", in fact.

However, when approached from this perspective, site-specific could also lead to *apparition* holding sway over appearance which, according to Marcel Duchamp, is part of the specificity of art and which seems more than necessary today in the face of the embedded hyperbole of the society of the spectacle (Paz 2000).

The current relevance of Duchamp's principle is evident from the predominance of spectacle in which "truth is a moment of the false" (Debord 2002), which is greatly exacerbated by the predominance of the virtual. This *Weltanschauung* extends to the entire urban condition, and site-specific interventions can only assume a heuristic value if they are able to translate into "formed form" not only the malaise of *living* reduced to mere *appearance,* but also the possibility of seeking meaning in something other than spectacle and the consolatory recomposition of the project tout court.

12.3 Forms of Precariousness: Places, Non-Places, Remains

The city as spectacle is the evolution of a mobile reality that, according to Benjamin, found expression in the arcades of Paris (Benjamin 2000). Deprived of utopia, Benjamin's arcades, which are nonetheless anthropological places, became Marc Augé's *non-lieux*, or non-places, the characteristic features of a city without relations (Augé 1993). Then, the non-places became "remains", sliding even further down the ladder of utopia within a nonwoven city, composed of what might be called an "unwoven fabric", in which the project and its ideals unravel due to a lack of interweaving that would give them meaning. This results in the multiplication of "remains" in the urban fabric, which corresponds to "improbable spaces". Places that do not invite speculation, and do not have the fascination of ruins or the dignity of industrial archaeology, but cannot strictly be considered waste. Recently discarded, these urban leftovers remain deadweights waiting for what will never be. They express the precariousness of the contemporary city's unwoven fabric, and at the same time, are one of its main characterizing elements. By intensifying its vocation for the short-lived pleasure of a continuous face-lift, the city of the spectacle reveals an extensive web of "wrinkles" and "dystonia" that undermine its consolatory aspect. The remains are too costly to remove. Assigning new permanent functions to them within an organic project is anti-economic. The new urban form is increasingly defined through the dispersion of leftovers, which rather than interrupting a complete planning discourse, are the words of a discourse that consists of interruptions, and as such, cannot be completed.

This effect would appear to be different from the one created by the change in paradigm, mainly characterized by "multiple" and "happy dispersions", defined by postmodern culture. More precisely, the urban leftovers were spawned by the crumbling of a palimpsest of collective meaning, caused by a kind of anthropology of insecurity, as Massimiliano Valerii has described it (Valerii 2019). In the global crisis, in an economic, political and social sense, but especially with regard to values, a state of widespread precariousness is becoming manifest, which is the outcome yet potentially also the beginning of new epistemologies. On the other hand, the transitory, the random, the unstable, the ephemeral and the indeterminate seem to be increasingly recurrent themes in scientific thought and reflection on culture in general, as well as in artistic research.

In this context, "place" acquires new characteristics, and consequently, requires us to rethink the notion of site-specific. Considered from a synchronic standpoint, its progressive evolution towards impermanence induced by the increased importance of the relational dimension is easier to understand; on the other hand, place demands that the concept of site-specific broaden its horizons to accommodate the nomadism of our times.

This hypothesis implies the introduction of a new frontality after postmodern obliqueness; a frontality based on the type of "altermodernity" referred to by Nicolas Bourriaud (Bourriaud 2014).

12.4 From Site-Specific to Time-Specific

What Bourriaud proposes is a wandering able to thrive on the disintegration of reality. For art making, he posits the principle of composition characterized by the transitory, defined as "journey composition". This results in a dynamic and impermanent vision of form—we would still say "formed form"—except that here formativity expands the dynamic datum according to the principle of wandering, instead of a process-oriented towards a goal.

By eliminating the goal, wandering is able to engage with the non-linearity of the itinerary. Indeed, the pulverization of the landscape offers endless opportunities to forge paths, each of which implies discovery. This transforms the precariousness of advancing into the potential construction of meaning. It is the need for "narrative" that emerges in the concept of journey composition, but it is a narrative *randonnée*, as the term is construed by Serres (1985); namely, as an epistemology founded on a zig-zag path, where the non-linearity of advancing has a heuristic value in itself. Through wandering "the journey-form puts linearity in crisis by injecting time in space and space in time" (Bourriaud 2014, op. cit.).

Due to the constant overlapping of space and time, the concept of site-specific shifts towards time-specific, since the space is the only space traversed at that particular moment of wandering. As a space of transit, the place becomes a momentary space and the work of art that comes into being in and from the place cannot but be momentary, unstable, volatile almost.

Hence, the transition from site- to time-specific infers fluctuating horizons capable of giving a cognitive value to a condition of permanent metastability. Time-specific in the form of *randonnée* cannot but be metastable, always about to strike a new equilibrium, and persistently positioned in an uninterrupted *during*. It is a choice that implies reduction, namely lightness, economy and flexibility of means and intervention. As with nomadic constructions, reduction is directly proportional to agility of movement: the more you subtract, the greater the potential mobility. Where subtraction means not only jettisoning "noble" or in any event definitive, materials, but also rejecting a conception of the artwork as a self-contained unmodifiable whole, oriented towards *duration* rather than *during*.

Space injected in time is the territory of the nomad and the nomad must travel light, on pain of permanence. Abandoned spaces pose for art the problem of identity and of materials capable of incorporating the temporary nature of the site through interventions that are equally temporary, marginal and fragile, but able to trigger processes of resignification through the place.

This opens up an area for experimentation where the relationship with urban leftover not only goes beyond the static vision of their reclamation as a container, but above all makes them an agent of poetics in the name of metastable equilibrium. Thus, they become real attempts at "negotiating" with the space, geared to "giving form to precariousness".

12.4.1 Giving Form to Precariousness: Two Case Histories

(1) *Dialogos*

Dialogos Part One.

Alessandro Castiglioni, Antonio Catelani, Sergio Breviario, Andy Boot, Ermanno Cristini, Giovanni Morbin, Giancarlo Norese, Goran Petercol, Fabio Sandri, Luca Scarabelli, Emily Speed, Alessandra Spranzi.

Curated by Alessandro Castiglioni and Ermanno Cristini.

Assab One, Milan, May 2011.

Dialogos Part Three.

Simona Barbera, Ermanno Cristini, Ronny Faber Dahl, Vladimir Havlik, Carlo Miele, Giancarlo Norese, Chiara Pergola, Luca Scarabelli, Una Szeemann & Bohdan Stehlik, Miki Tallone.

Curated by Ermanno Cristini and Giancarlo Norese.

Vitrína Deniska, Denisova 5, Olomouc (CZ); Galerie Caesar, Horní náměstí 583 (radnice), Olomouc (CZ), May–June 2017.

Dialogos is a project launched in 2010, which has generated various exhibitions, including two in particular in abandoned spaces used for impromptu artistic interventions. One of these spaces is Assab One in Milan, a former printshop housed in a building with a surface area of around 3,000 sqm; the other is a small shop-window in disuse, which functions as a project space, in the centre of the town of Olomouc in the Czech Republic.

The two venues are very different places with regard to their history, size, physical characteristics and identity, yet are both sites "thrown into disuse". Vitrína Deniska is a microspace in an abandoned building that does not even have the fascination of a "ruin". Whereas Assab One is a characteristic and atmospheric site but does not possess the dignity of industrial archaeology in the strict sense.

In both cases, *Dialogos* has given rise to experiments that use the space in a "time-specific" rather than "site-specific" way, due to the transitory nature of its approach.

Dialogos is an exhibition project midway between performance and installation. It sprang from the possibility of considering the space as a formula of time, first a *when* and then a *where*, since it is negotiated and transformed by relation.

The *Diagolos* formula is simple: a dozen artists interact through their works, for an unspecified and extended period—over a year— in a document shared online. One of the artists starts the ball rolling, throwing a "pebble in the pond" to which the others respond, and the document takes the form of a conversation through images, texts, sounds and so forth: works and semifinished works. At a certain point, the contents of the document are projected in a space and reworked through the stimuli offered by the place, understood not only as a testing ground for dialogue, but also as a term of its language.

Dialogos introduces the concept of display as negotiation. One artist responds to another, which results in display devoid of the static quality of a definitive staging. On the contrary, what is staged in the exhibition actually displays what is not seen, its darkness, the silence that takes on the meaning between the "words". Behind the visible artefacts, we can sense stories. And, paradoxically, what is absent is more meaningful than what is represented.

As if in a chess game, the artistic intervention and its sign, the object, take on a form of reversible communication: between artist and artist, artist and work, work and space, space and viewer, viewer and space, space and work, work and artist, artist and artist. The mechanism created by the project is thus composed of actions, responses and responses to those responses, all articulated in a fugue of formal and always temporary solutions.

The leitmotiv of the 2011 edition of *Dialogos* held at Assab One was light. The work of the artists had begun around a year earlier and explored the relationship between light and shade, and the distance that separates, and at the same time, unites them. The large industrial windows are an important feature of the Assab building, especially on the upper floor that originally housed the photolithographic department of the company. At different times of day, the physiognomy of the space changes due to the light effects.

One of the participating artists, Fabio Sandri, created a device to capture, ideally, the action of light in relation to the passing of time. It was a rudimentary camera obscura made from a cardboard box and supported by a very light metal structure (Photo 12.1). Next to it was a pile of sheets of photosensitive paper, separated by horizontal pieces of card, placed one on top of the other. Every morning a photosensitive sheet was inserted in the back of the camera obscura and left inside all day to ideally "capture" the light variations in the space. The next morning the sheet was extracted and placed on the floor to form another pile next to the first one.

As the days passed, the initial pile diminished as the second pile of sheets on which the changes were recorded, grew proportionately. The decreasing and increasing piles of paper suggested the passing of time, giving rise to a transitory environmental work that did not represent light but *presented* its becoming as the generative element of a space. In response to this work by Sandri, Goran Petercol focused on artificial lighting. Taking a table lamp found on the premises, he removed and distanced the light bulb from its socket in order to jolt the viewer's perception: creating darkness where the light was usually created and light where there was normally shade. The distance was marked by the line described by the electric wire that both linked and contrasted the two elements, which were separated by a plinth—also found on the premises—thus deprived of its usual pedestal function (Photo 12.2).

Antonio Catelani, Alessandra Spranzi, Luca Scarabelli, Ermanno Cristini and Emily Speed, on the other hand, "shed light on" the fundamental stages of their interventions through the use of objects, considered as *wreckage.* Dialogue is a journey, and by definition, inconclusive, thus implying running aground. Wreckage is what remains after a shipwreck; it is waiting to be brought ashore, to be lined up on the

Photo 12.1 *Dialogos Part One*, 2011, Assab One, Milan, ph. Ermanno Cristini

beach and classified. In this case, the space functioned as a categorizing table and the elements were "installed" on the floor by creating a list and a condition of waiting, of suspended time through a hypothetical classification.

The only compositional aspect of the formal outcome was that of placing the elements side by side to establish possible links between them. This gave rise to grids of meaning, which were arranged in the space like a "blackboard", becoming the field of a potential narrative (Photo 12.3).

Light and shade, volumes and voids, vertical and horizontal, this was how the architecture of the place was activated by micro-interventions that occupied it to establish it and vacate it immediately afterwards, with the aim of researching its identity always returning to square one.

In 2017, the third edition of *Dialogos* was held in a space that was, by contrast, extremely small, the Vitrína Deniska in the Czech Republic. The exhibition was

Photo 12.2 *Dialogos Part One*, 2011, Assab One, Milan, ph. Ermanno Cristini

staged, in parallel, at a project space in the centre of the town of Olomouch and in the shop-window that usually has an independent calendar and curatorial team. In this case, the dialogue between artists also took the form of a dialogue between the spaces. Luca Scarabelli applied an adhesive sheet of reflecting material with two central holes to the glass of the Vitrína Deniska.

The intervention enclosed the "focus" of the window by positioning it between "looking in" and "looking out". Carlo Miele retrieved a fragment of plaster that had detached from the frame and treated it like a patch of colour, placing it at the base of the window itself, while Chiara Pergola amplified the chromatic action with a length of mirror which, at the same time, related to the reflecting surface of the adhesive material in the centre of the space (Photo 12.4).

Apart from the specific character of the formal solutions adopted, it is important to note that here a residual place was able to take on new life, the temporariness of its resignification becoming an element of strength. A strength that was directly proportional to the lightness of the interventions, which, first and foremost, made it a practical and hence accessible space. Lightness is also understood as the "subtraction of weight" that Italo Calvino in his *Six Memos for the Next Millennium* (Calvino 1988) sees as the principle of "thoughtfulness" suitable for giving form to the ephemeral nature of reality. "Thoughtfulness" is the form of temporal slowing down in which elements of a minimal physicality reemerge. Manifestations of a dormant humanity that can provide an opportunity for the regeneration also of places.

Photo 12.3 *Dialogos Part One*, 2011, Assab One, Milan, ph. Ermanno Cristini

(2) *Soprasotto*

Above and Below. Disequilibriums of Vision.

From Disequilibrating Systems by Ugo La Pietra.

Alessia Armeni, Umberto Cavenago, Elena Ceci e Ana Victoria Bruno, Ermanno Cristini, Silvia Hell, Sergio Limonta, Corrado Levi, Joykix, Michele Lombardelli, Monica Mazzone, Yari Miele, Bruno Muzzolini, Giancarlo Norese, Luca Scarabelli, Miki Tallone, Sophie Usunier.

Photo 12.4 *Dialogos Part Three*, 2017, Vitrína Deniska and Galerie Caesar, Olomouc (CZ), ph. Ermanno Cristini

Curated by Ermanno Cristini and Luca Scarabelli with critical contribution by Rossella Moratto.

Organisation by Elena Ceci and Ana Victoria Bruno.

Former supermarket Bottegone, Sesto Calende, June–July 2016.

Another project concerned with the shift from site-specific to time-specific was realized in a former supermarket in disuse at Sesto Calende, in the province of Varese.

Constructed in the 1980s, the building had no significant architectural features and had been urban leftovers for some time (Photo 12.5).

The exhibition *Soprasotto. Disequilibri della visione* (Above and Below. Disequilibriums of Vision) was created to coincide with the anthological show of Ugo La Pietra at the Museo MAGA in Gallarate (Meneguzzo 2016), since it was indirectly and tangentially related to his *Sistemi disequilibranti* (Disequilibrating Systems) of the 1970s (La Pietra 1970). Here, the aim was to experiment with the temporary resignification of the site by giving expression to marginal and insignificant elements of the building, seen from an unusual viewpoint.

In the catalogue we read.

"This exhibition is designed around the notions of above and below, since it unfolds entirely on the ceiling and the floor, with the sole exception of a divergence of 70 cm below the ceiling and of 70 cm above the floor (the exhibition space).

The idea is to explore "another" landscape, with works that are below eye-level but above the floor, and below the ceiling but above eye-level. *Above* and *below* describe the space to possess it, touch it and orientate us, while also seeming to negate the anthropological depth of the horizon we are confronted with daily. High and low have a powerful symbolic value and the same binary logic which, according to Lévi-Strauss, helps us to order the world. A geometric order explored by the senses.

Photo 12.5 *Soprasotto*, 2016, Former supermarket Bottegone, Sesto Calende, ph. Miriam Broggini

But if we reverse the terms and levels, the opposing pair creates confusion and disequilibrium, through which deceptions and new ways of seeing emerge. In the real sense, as an exercise in observing, and in the metaphorical sense, as a place in which certainties waver. A kind of telluric movement, which is also the condition of utopia, a healthy exercise in all those cases where we find ourselves having to explore the limits of one world to create a new one" (Cristini 2016).

In using the building, we accepted its status quo, including the absence of electric light, which had not worked for some time. Since the perimetral glass surfaces were of considerable size, the illumination was entrusted entirely to daylight, meaning that the works and the exhibition as a whole varied in appearance at the different times of day, emerging in the morning and slowly disappearing in the evening. At the same time, the hierarchy of the elements of the space changed according to the time of day and always created different landscapes. Within this temporal "narrative" the artist's works were arranged in such a way that the viewing experience was characterized by a continuous looking up and down.

Umberto Cavenago attached two small shelves to one of the double-glazed windows using suction cups, at a height of 70 cm from the floor. A pile of out of date coins and another of buttons, which belonged to a family collection, were arranged meticulously on the respective shelves (Photo 12.6). With the passing of time and the action of the varying temperatures on the glass, the suction cups came unstuck and the two ordered piles fell and got mixed up, destroying their rigid classification and creating a new composition of unforeseen and unpredictable materials on the floor (Photo 12.7). Giancarlo Norese arranged the full-length photographic

Photo 12.6 *Soprasotto*, 2016, Former supermarket Bottegone, Sesto Calende, ph. Luca Scarabelli

Photo 12.7 *Soprasotto*, 2016, Former supermarket Bottegone, Sesto Calende, ph. Miriam Broggini

portrait of two people by dividing the print horizontally in the centre and placing the two halves, the one with the feet and the other with the heads, respectively, on the lower part of the wall, in contact with the floor, and on the upper part, in contact with the ceiling. To ideally recompose the figure, the eye had to constantly move up and down (Photo 12.8).

Alessia Armeni exploited the light of a window to build a work around colour, fixing a terrestrial globe with a particular colour scale to the ceiling and a humming top with the same colour scale to the floor. The changing light gradually modified the chromatic impact of the two elements, indirectly enhancing the space that separated them and giving them new potential meanings (Photo 12.9). Ermanno Cristini utilized some pallets that he found on the premises as bases on which to arrange a series of makeup mirrors with a wood frame—the same material as the pallet—facing upwards. The ceiling was reflected in the mirrors, overturning and reversing the relationship between above and below (Photos 12.10 and 12.11). Luca Scarabelli emphasized the horizon of the exhibition, by positioning a broom, also found onsite, in the centre of the space, with the handle cut down to a height of 70 cm from the ground (Photo 12.12). Corrado Levi gave a cello performance, transforming the former fishmonger's space, characterized by walls faced with small ceramic tiles, into a stage. Monica Mazzone created aluminium structures that evoked the industrial materials of the building and its fittings (Photo 12.13). Miki Tallone dismantled most of the ceiling lights then aligned them on the floor to construct a minimalist geometric structure, which was enlivened by the disorienting experience of seeing on the ground what was normally above.

Photo 12.8 *Soprasotto*, 2016, Former supermarket Bottegone, Sesto Calende, ph. Miriam Broggini

Thanks to these and other interventions, visiting the exhibition became a novel way of experiencing the space, whose critical aspects became a language. Observed by a "foreign eye", stimulated by the artists' work, the actual character of the place emerged. Its memory emerged, through the aesthetic use of some of the abandoned elements, like the ceiling lamps, pallets, a trolley and so forth. Its volume emerged through the enhancing daylight that gave substance to the voids of the windows and sculpted the volumes of the walls. Its materials emerged through the works that reintroduced sheet metal, aluminium and the like. Its identity emerged, caught in the act of a possible transformation.

Photo 12.9 *Soprasotto*, 2016, Former supermarket Bottegone, Sesto Calende, ph. Miriam Broggini

Photo 12.10 *Soprasotto*, 2016, Former supermarket Bottegone, Sesto Calende, ph. Miriam Broggini

Photo 12.11 *Soprasotto*, 2016, Former supermarket Bottegone, Sesto Calende, ph. Miriam Broggini

Photo 12.12 *Soprasotto*, 2016, Former supermarket Bottegone, Sesto Calende, ph. Miriam Broggini

Photo 12.13 *Soprasotto*, 2016, Former supermarket Bottegone, Sesto Calende, ph. Miriam Broggini

References

Augé M (1993) Nonluoghi. Introduzione a una antropologia della surmodernità. Elèuthera, Milan
Barbero LM, Pola F (2010) L'Attico di Fabio Sargentini 1966–1978, exhibition catalogue at MACRO, Rom, Electa, Milan
Benjamin W (2000) I «passages» di Parigi. Einaudi, Turin
Bourriaud N (2010) Estetica relazionale. Postmedia, Milan
Bourriaud N (2014) I radicanti. Postmedia, Milan
Cage J (2013) Al di là della musica. Mimesis, Milan-Udine
Calvino I (1988) Six memos for the next Millennium. Harvard University Press, Cambridge MA
Cristini E (2016) Soprasotto. Disequilibri della visione, exhibition catalogue, Cesare da Sesto Editons, Sesto Calende
Debord G (2002) La società dello spettacolo. Massari Editore, Bolsena
Doherty C (ed) (2009) Situation. MIT Press Ltd, Cambridge
Gruppo μ (2018) Semiotica e retorica della cornice. In: La cornice. Storia, teorie, testi, a.c. of Daniela Ferrari and Andrea Pinotti, Johan & Levi, Monza
Guida C (2012) Spactial Practices. Funzione pubblica e politica dell'arte nella società delle reti. Franco Angeli, Milan
Havranek et al (2006) Jiri Kovanda: actions and installations 1976–2006. JRPEditions, Geneva
Kwon M (2020) Un luogo dopo l'altro. Arte site-specific e identità localizzativa. Postmedia Book, Milan
La Pietra U (1970) Il sistema disequilibrante. Galleria Toselli Editions, Milan
Meneguzzo M (2016) Ugo La Pietra. Il segno randomico. Opere e ricerche 1958/2016, exhibition catalogue "Abitare è essere ovunque a casa propria", Silvana Editoriale, Milan
Mock J-Y (ed) (1983) Yves Klein, exhibition catalogue, Centre George Pompidou. Musée National d'Art Moderne, Paris
O'Doherty B (2012) Inside the white cube. L'ideologia dello spazio espositivo. Johan & Levi, Monza
Palazzoli D (1967) L'artificio non come retorica ma come logica del costituibile. In: Con temp l'azione, exhibition catalogue, galleries edition Il Punto, Sperone, Christian Stein, Turin
Paolini G (2006) Quattro passi. Nel museo senza muse. Einaudi, Turin
Pareyson L (1974) Estetica. Teoria della formatività. Sansoni, Firenze
Paz O (2000) Apparenza nuda. L'opera di Macerl Duchamp. Abscondita, Milano
Pissarro and Hunter College Art Gallery (2016) Critical gestures & contested spaces: art & politics in 1960s France. Hunter College, the City University of New York
Ratcliff C (1983) Andy Warhol. Abbeville Press, New York
Rose B (1965) ABC art. In: Art in America
Serres M (1985) Les cinq sens. Grasset, Parigi
Valerii M (2019) La notte di un'epoca. Salani, Milano

Chapter 13
The Polysemantic Nature of Music. A Hermeneutical Perspective

Pier Francesco Forlenza and Michael Giovanni Jennings

Abstract In this short essay, we will discuss how the concept of the *leftover* can be applied with interesting results in music. In fact, we find that the idea of the *leftover* underlies musical discourse in three ways, which are all intertwined with one another, while maintaining their own specificity. These three declinations define the *leftover* as a synonym of *surplus*, *renewal,* and *imperfection*. Firstly, the *leftover* can be intended not in a pejorative acceptation, as common sense would suggest, but rather in a positive and constructive one. In this regard, the *leftover* is seen as a *surplus* or *abundance*. The development of this idea starts off with some brief reflections on the relation between music and language and on the impossibility of an "equal" translation of the former to the latter. We hold that by trying to put music into words, what is left behind is an *excess* of meaning. Our second interpretation of the concept—*leftover* as *renewal*—is a partial consequence of the first abovementioned connotation, inasmuch as every reutilization of something implies that the very object of renewal possesses an excess of unexpressed possibilities. In this section, we show how two composers, A. Schönberg with his *Suite* Op. 25, and M. Ravel with *La Valse*, have utilized ancient forms—baroque dances, and waltz, respectively—to contain extremely novel musical contents. In addition, we hint at how the evolution of the *sonata form* can be interpreted in light of this second interpretation of the *leftover* and discuss how a *renewal* can happen at an even more abstract level, by comparing the harmonic structures of two of the most famous cycles in classical music, i.e., J.S. Bach's *Well-Tempered Clavier* and F. Chopin's *Preludes*. Finally, the third reading relates the *leftover* to the etymological meaning of *imperfection*, i.e. something which is left *unfinished*. From this standpoint, by ideally recomposing the history of improvisation, we trace its origin in the baroque practice of *figured bass*, showing how it contains principles common to contemporary jazz performance habits. To conclude, we come back to Bach's *Preludes and Fugues*, readapted by pianist Brad Mehldau, as a clear example of all of these three acceptations of the *leftover*.

P. F. Forlenza (✉) · M. G. Jennings
Conservatorio di Musica G. Verdi, Milan, Italy

L. Crespi (ed.), *Design of the Unfinished*, The Urban Book Series,
https://doi.org/10.1007/978-3-030-73457-2_13

Keywords Ambiguity · Estrangement · Expressivity · Figured bass · Improvisation · Remnant · Musical architecture · Musical narration · Sonata Form

13.1 Leftover as Surplus

Our journey through the multifaceted nature of the *leftover* begins with a reflection on the relationship between music and language. These two essential elements of human nature have indeed attracted and engaged many thinkers in vibrant confrontations, which have accompanied the development of human thought throughout the ages, from Plato to today's most eclectic intellectuals. The ways in which music and language relate to one another are of course infinitely numerous, so what interests us is only a small part of their complex interaction, specifically their shared capacity to convey meaning and sense, to be expressive of something and thus able to communicate it. This common feature allows us to reflect on one in terms of the other.

In this regard, it is illuminating to bring to mind two apparently opposite lines of thought that show how rich and ambiguous musical and linguistic intertwinings can be and that help us to shed light on our first acceptation of the *remnant*. The first comes from a letter, which Mendelssohn Bartholdy (1842) sent to a friend, claiming that music is able to convey and arouse thoughts that are at once too precise and accurate to be put into words. The second is instead embodied in Jankélévitch's (1961) famous idea which holds that music is ineffable, inexpressive, and ultimately unable to mean anything. This polarization between precision on the one hand and ineffability on the other raises the issue of how music relates to language and how it can be translated into it, if even possible.

In the first case, the claim underlines how language fails to convey what music expresses, precisely because it is too specific to be understood and explained in its full semantic significance: a translation into words would compromise its comprehension, inasmuch as the phenomenal character of our musical experience—would be, so to speak, left out. It is indeed difficult to deny the idea that each piece of music is able to express something, which is specific to that particular piece and that is not shared by any other composition. Ridley's (1995) words can better capture what we are trying to suggest:

> The 'marcia funebre' from the 'Eroica' Symphony, for instance, is expressive of an affective state which is quite unlike that of which the funeral march from Chopin's Piano Sonata No. 2 is expressive; and neither affective state seems similar to the state conveyed by 'Siegfried's Funeral March' in Götterdämmerung. Each of these works is clearly expressive of some sort of funereal sentiment. But when asked what are the differences among those sentiments, most of us will be at a loss for words—or at least at a loss for words which we feel are the right ones.

From this consideration, what comes to mind is Wittgenstein's (1953) investigation of the process of understanding a proposition. In his view, we are able to comprehend a given sentence in such a way that it can be either substituted by another with its same meaning or understood as unsubstitutable. The implication is that the meaning of a proposition can sometimes be shared by multiple ones, and that other times it can be extremely specific to a particular proposition, to its precise word order, as, for instance, in poetry. It is now easy to see how this can be applied to music as well and to the previous quote. Indeed, this is exactly what Mendelssohn refers to as the precision of music, insofar as each funereal sentiment is certainly shared by many pieces, but never the same: each composition has its own type of sentiment, intimately tied to the specificity of its structure and perceptual unfolding, which is only perceivable by experiencing it and not by describing it. Thus, it is in this sense that music can be more precise than language.

If it is fundamentally possible to talk about music as having an accurate and precise expressivity we must always keep in mind that it can truly be too precise for words, just as Schumann demonstrated by playing a piece over again when asked what it meant to him. This uniqueness of meaning suggests that the content of music is too big to be confined to a linguistic container, and clearly produces an excess of sense that is crucial for our understanding of the *remnant* as a form of *surplus*.

Returning to the twofold distinction made at the beginning of this section, it is now time to look into Jankélévitch's side of the coin. If Mendelssohn has shown that music is too specific to be verbally understood in the entirety of its significance, Jankélévitch's point of view seems, instead, to be quite the opposite at a first glance. The claim he holds is that, ultimately, music is in itself ineffable and has no meaning. Interestingly though, the path to this conclusion shares many similarities with what Mendelssohn had expressed a century before him.

A linguistic approach to music gets in the way of music itself: it imposes a constant reference to discursive categories and to its developmental structure—introduction, content enrichment, climax, conclusion—, which are imposed by our conversational habits, and ascribes value to it in virtue of its similarities with language. From this point of view, for instance, to say that music is dialogical is fruitless, because its dialogical characteristics are an attribution of our experience. As such, the repetition of a motif, first played by strings, then recalled by woodwinds, does not constitute a dialog that would by any means resemble an actual conversation between humans (and truly, who would repeat what one has just heard?). This is to show that if music has meaning and if it can be expressive of something, it is certainly not because of its linguistic capabilities.

Nonetheless, it is hard not to see in music a narrative structure, informed by many factors, such as dynamics, pitch, tempo, tone, character, and so on. Of course, this appears immediately objectionable, given that narration is too, in a way, a superimposed causal interpretation of certain events (Damasio 1999)—i.e., sounds in this particular case. A response to this objection, though, is that, as shown by Damasio himself, our narrative instinct precedes language by being its very prerequisite: after all, there can be no narration if there is no narrative to be verbalized. Going back to our main focus, Jankélévitch's idea is that music does have a narrative quality that is simply not of a linguistic type: music can describe, portray, and evoke, but only, as he says, "*en gros*". It can express the "*sense of sense*". The titles of Debussy's Preludes, for instance, enclose a sense that is far bigger than the determinacy of its denomination: music can thus express the infinite possibilities, suggestions, and in short, the imaginative horizons contained therein.

Jankélévitch does not, however, explicitly write about music with no descriptive title, but we can surely stretch these intuitions to the domain of non-programmatic music, although he would not approve, for reasons that are outside our purview here. In the end, our interest is to freely reflect on his perspective, not to offer an accurate account of it. All music, after all, offers an infinite spectrum of imaginative possibilities without excluding one or another. All of these interpretations coexist at the same moment, in virtue of being constituent of what defines a particular piece of music as *that* piece of music. This evanescent quality of hermeneutical multiplicity is the core of music's ineffability and thus of its inexpressibility. The freedom of choice that music grants us upon listening to it feeds on its boundless wealth of sense: the "*inexpressive espressivo*" epitomizes this quality in its full semantic range and associative connections. Seen this way, Jankélévitch's assertions hopefully make more sense now: music does not mean anything, precisely because it can mean everything.

It is now clear how our first acceptation of the *remnant* can be justified. The value of music resides in its *overabundance* and *surplus* of meaning that becomes even more apparent when analyzed in relation to language, as if it were a litmus paper. Both lines of thought, with quite different premises, arrive at the same conclusion. Music generates a *remnant* of sense that makes our experience of it treasured and meaningful.

13.2 Leftover as Renewal

Moving on to considering the term *leftover* in our second acceptation—that of *renewal*—we will analyze specific cases in which this revitalizing operation takes place on a formal and abstract level, without discussing those cases of reutilization of compositions per se; by talking about re-elaborations, transcriptions, variations, and paraphrases, it would be necessary to extend our considerations to virtually all music, or to arbitrarily select examples with no true connection to this analysis.

(a) *Schönberg and Ravel: fast and loose with musical tradition*

One of music history's most emblematic examples is Arnold Schönberg's use of traditional dances, such as the Gavotte, Musette, Minuet, and Gigue, concurrently with the development of his dodecaphonic compositional method. This Copernican revolution, which abolishes the typical tonal hierarchies between sounds and reunites all twelve notes of the chromatic scale around the tone row's democratic order, was achieved while retrospectively drawing on antiquated and conventional seventeenth and eighteenth century dance forms. Conceivably, this was intended to reassure the listener, to temper the impact of the complete novel elements of this new musical vocabulary. The unusual relationships between pitches, the prevalence of unorthodox intervals, the annihilation of the melody-harmony distinction, in favor of a Cubist equivalence between a horizontal melodic level and a vertical harmonical layer, are all diluted and made more accessible as a consequence of being placed in a familiar sonic landscape, at least rhythmically.

If the *Suite* Op. 25 or the last of the five pieces (a waltz), which comprise Op. 23, can theoretically be danced on, it follows that something innovative, yet familiar and already heard is woven into it. The rhythmical heritage, derived from those dances, helps to temper this music's apparent lack of direction, due to its being regulated by a principle of infinite variation, and the absence of thematical points of reference. It is evident that this process of recognition is, essentially, a guarantee of sense for the listener.

It might be opportune here to recall that Pierre Boulez, exponent of a radical tendency, which stemmed much more from Webern than from Schönberg, interpreted these principles in a completely different and negative key. According to the French composer and conductor, Schönberg's limit was having failed to draw the full consequences of his dodecaphonic discovery: the new system of sound organization should have extended to the categorization of note value, type of sound attack, indeed to all sonic parameters, and should have ultimately been part of a more integral formal and structural *renewal*. Instead, the Austrian composer, as if frightened by his own innovations, attempted to force this new language to coexist with completely incompatible preclassical forms. Moreover, as Boulez (1966) points out, having maintained tonal "infrastructures" impeded a more complete stabilization of the new system. Nonetheless, in order to correctly frame Schönberg's operation through the *reutilization* of ancient dance forms, we must keep two things in mind. Firstly, if we were to accept Boulez critique, we would end up reproaching Schönberg for not having written Boulez' music thirty years earlier, which would be antihistorical, if not outright absurd. Secondly, it must be said that Schönberg himself did not aspire to be considered a revolutionary and that he insistently and provocatively asserted the continuity of his poetics with that of the past, explicitly refusing, as written in his *Harmonienlehre* (1911) the label of atonal composer, "*a term from which I must keep distance, as I am a musician and have nothing to do with atonality. This word can only denote something which does not correspond to the nature of sound*".

In conclusion, we can say that hearing these sixteenth century dances in Schönberg's music produces a strong sense of estrangement. The dodecaphonic language allows these forms to appear as if innervated by an unusual substance that transmutes them into genetically modified organisms, becoming a sort of transminuet or transgavotte, a metagigue or metawaltze.

It is interesting to note that, almost around the same period in which Schönberg wrote his first dodecaphonic piece—the closing waltz of the *Klavierstücke* Op. 23—Maurice Ravel published his *poème chorégraphique*, unambiguously titled *La Valse*. A waltz, then, once again. Composing one at the beginning of the '20s, right after WWI, in the full modernist spirit of iconoclastic fury that characterized the European artistic scene, could have certainly been seen as a sign of folly or naïveté—especially in the case of a prominent composer and self-aware artist and not of an aspiring musician. This composition, which could have appeared as a plain stylistic exercise or a disengaged intellectual frolic just a decade before (“*le plaisir délicieux et toujours nouveau d’une occupation inutile*”, as written on the epigraph of the *Valses nobles et séntimentales* in 1911), becomes instead a completely anachronistic gesture in the aftermath of the Great War. Yet, what Ravel actually accomplishes is incredibly surprising.

For quite some time the composer had in fact entertained the idea of writing an homage to J. S. Strauss II’s dazzling ball dance: he had even thought of entitling it *Wien*, were it not for the impact of the brutal conflict between France and the Habsburg Empire, which led him to change his mind on this project. So, it was only at the end of 1919 that Ravel resumed the work that would come to light the following year, in a triple version for solo piano, for two pianos, and for orchestra. There is truly nothing reassuring in this *Valse* that, on the contrary, opens with magmatic menace and continues with crazed obsession, enticing, insolent, restless, and unnerving. The homage to an obsolete genre and the quotation of its *clichés* become so sophisticated that the iconic nineteenth century dance metamorphizes into something completely different. The typical waltz ethos is only evoked, often eluded, negated, and denatured to become a primordial ritual dance. It is a composition that possesses, recalling Jankélévitch, great ambiguity: on the one hand, it appears as the most refined waltz ever written, as the final celebration of a world and lifestyle perhaps longed for. On the other, with its extreme harmonic complexity, its kaleidoscopic variety of color and timbre, and its transfiguration of the rhythmic pulse into ancestral kinetic energy, this waltz resembles more a new *Sacre du Printemps,* than the bourgeois dance from which it derives. Its saturated colors suggest that it could be a synesthetic conclusion, an imaginary soundtrack to Matisse’s *Danse.*

Here, as well as in other occasions, Ravel manages to create an extremely personal and advanced harmonic syntax, made of seducing sonorities and subtle balance. Its influence reaches afar, so much that jazz itself, to which we usually ascribe the highest level of harmonical experimentation, owes a lot to it, only reaching its levels of complexity a few decades later. Nonetheless, Ravel’s prophetic music does not show any rupture with tradition, always establishing itself on a line of continuity with the past. As a magician though, the more Ravel remains attached to his inheritance, the more he negates it. As in this *Valse,* in which the *remnant* is the pretext to start a process destined to overturn its premises.

(b) *Schumann, Liszt, and structural freedom*

Going now back in time, from one musical discovery to another, we would like to discuss how the acceptation of *remnant* as *renewal* can also be seen in the development of one of Western Music’s greatest inventions. We are talking about perhaps the most important formal principle, which, conceived during the Preclassical period, survived well into the twentieth century. The *sonata form* is certainly only one episode in a complex chronicle of styles and principles of musical organization, but it is a pivotal and decisive one, as it provided the structural medium for a great deal of the most profound musical thought until half of the nineteenth century, and continued to be prominently present in the works of many other composers.

Briefly laid out, the *sonata form* is a structural imprint that has been normally associated with solo or instrumental music, in particular, to first large scale movements of compositional cycles, as a Quartet, a Trio or a Symphony. Nonetheless, this, just like anything else written about the *sonata form*, is not in any way prescriptive, given that as in any existing instance of a formal principle, its method vastly differs from its possible applications. We must never forget that its practical use and development preceded its conceptual denotation. What makes the *sonata form* an incredible feat, though, is the fact that it manages to confer a psychological character to certain formal elements, creating a narrative arch throughout its unfolding. Its basic elements are usually three: exposition, development, and recapitulation, during which the thematic material becomes, respectively, stated, expanded, and reproposed. Thus, its ternary structure is often apparent at first sight: the three parts are interrelated not in terms of basic structure, but by purely thematic, melodic or character contrast. Actually, the three parts of sonata form originated out of the binary, or two-part form conventionally used in the seventeenth and early eighteenth century music. In this form, the structure depended not simply on a thematic interrelationship, but also on a tonal one. In this way, the first part flows directly into the second, by ending in the key in which the latter begins; vice versa, the second proceeds from the new key back to the original one, in which the piece ends. The circle is closed, the journey concluded, as the second part completes the first. It is certainly striking to observe how in certain baroque pieces the symmetry of this formal duality becomes more and so often experimentally stretched, by greatly expanding their second part, enriching it with more material, insight, and depth, as in J.S. Bach's *Sarabande* from the *D Major Partita* BWV 828.

We can trace and superimpose this binary form to the ternary structure of the *sonata form*: the exposition corresponds to the first part, while the development, the moment of expansion, and its subsequent recapitulation correlate to the second. The exposition moves from the original tonality to a new one, whereas the development roams through multiple ones and the recapitulation returns to the original key. Just as in the binary form, the harmonic motion from and to the original key is maintained, although with much more complexity. This organic relationship between parts defines the *sonata form* as a far more elaborate structure, where the crucial driving force is the necessity of contrasting and juxtaposing, whether starkly or smoothly, musical statements or ideas that become dialectically handled, combined, dismantled, and finally recomposed. As suggested, though, the biggest contrast of all happens on a tonal level, which confers a sense of direction and narration to the structure and its components.

The ways in which the *sonata form* evolved from the preclassical period are quite numerous, and therefore, both successful and unsatisfactory. Two positive and interesting paths that it took are certainly exemplified by R. Schumann's *Fantasie* Op. 17 and by F. Liszt's *Sonata*. In these two outstanding compositions, pillars of Romantic literature, their authors invented and deployed a strikingly original method to transmit a natural formal fluidity, in which themes, melodic lines, structural elements transmute without ever changing from their primordial essence. This interplay of perspectives is typical of these two composers, who have strived to create a poetics based on free, fantastic forms, each according to their own specific personality. Schumann's particular tendencies of writing freely reinterpreted, unsettling, and fragmented music become exacerbated in Op. 17, just as Liszt's hypertrophic titanism and expressive grandeur overflow from his *Sonata*. These two works also share a peculiar relationship in music's history, as they are reciprocally dedicated by the two composers to each other, despite a 14-year difference in their genesis. What interests us about these two compositions is the way in which, despite the apparent rupture with the past tradition, the *sonata form* and its principles still underlie their large scope and structure. The main point of continuity is formal cohesion, which is not obtained by applying more or less rigid schemes, but rather by assuring a flowing fluency of musical narration. As previously seen, thematic and harmonical development was the core of the success of *sonata form*, yet Schumann manages to elevate those aspects to unprecedented heights.

The *Fantasie* is a work in three movements, each very peculiar and tied to tradition and innovation. The first movement already shows numerous structural anomalies. The starkly contrasting themes normally present in a *Sonata form* are substituted by extended, lyrical melodies, occasionally interrupted by a short secondary subject and some transitional sections. The architecture is filled with dramatic, temporal, and stylistic changes and with subtle turns, where themes suddenly dissolve or unexpectedly surface. In fact, the thematic material undergoes development immediately, causing a strong sense of formal ambiguity. The first melodic line, of quite some length, opens the piece with great authority and declamation: indeed, this will basically be all of the first movement's thematic material, wittingly varied and dissected, in order to grand the freshness of novelty, but also a solid sense of continuity. The melody thus serves as an embryonal and primordial element for all of the piece, *renewing* the tradition of the *sonata form* into a more compressed, yet freer manner. This is also clear from the opening, which emphasizes the dominant side of the tonality—incredibly, the tonic key of C major is not decisively established in root position until the very last chord. The choice of C Major is in itself another element of *renewal*. Typically thought of as either a solemn or simple key, Schumann utilizes it by blurring its luminous color: the *Fantasie* begins with a sonic effect, an incomprehensible sequence of fast notes which progressively add harmonic tension to the, on the opposite, clear melodic line above. The result is that C Major is no longer neither solemn nor innocent, but oneiric. The conventional development section, normally marked by inventive modulation and motivic manipulation, is instead substituted by a set of variations with various episodes, frequent intrusions, or quotes from other sections of the movement. This long central digression is unexpectedly in C Minor, giving a sense of unsettling and disheartening bewilderment to the narrative flow. The recapitulation instead, begins parallel to a point well in the exposition, but with the *incorrect* pitch, A-flat, replacing A as the tone of melodic outset. The coda, written as a structural compensation for the mangled introduction, is the full account of a musical quotation by Beethoven—from the song cycle *An die Ferne Geliebte*—only alluded to during the development, as if too important to be fully unveiled until the very end.

The *Fantasie* as a whole, with its two other movements, repeats itself structurally, as the second and third sections are built in a similar fashion and often with closely related material to that heard in the opening movement. The themes of the *Fantasie* are treated in such an elastic way that only their elemental features are carried on, loosely securing the structure, and creating a dreamlike progression that always gravitates around known and heard ideas. Schumann's choice of material is carefully deliberate and upholds unity while granting great fluidity and ambivalence.

From a certain standpoint, the same can be said about Liszt's *Sonata*, although he was surely not as inclined as Schumann to implicitly maintain a formal heritage in his compositions, as if almost fearing that a rigid architecture could impede a complete and free self-expression. Unknowingly or not, he did so anyway. It is striking that in his immense array of compositions, ranging from *Symphonic Poems,* to his musical travel journals, the *Années de pèlerinage,* or to different types of *Etudes* and transcriptions, the *Sonata* emerges as a solitary experiment. Its uniqueness though is also tied to Liszt's grandiosity and to the underlying presence of an extramusical plan, which follows *Faust's* sequence of events and most importantly to its architectural polyvalent character.

This work appears, in fact, at the first glance as an enormous *Sonata form* movement, in its traditional double theme and tripartite articulation; at a more careful observation, though, the agogic transformations, which disclose from the development of all multiple themes, allow to interpret this *Sonata* as a composition in four movements—*Allegro*, *Adagio* section, *Scherzo*, and *Finale*—which seamlessly follow one another. As further proof of formal polyvalence, pinpointing the beginning of recapitulation is ambiguous, given that it could be either placed at the main theme's recurrence, in B minor, the home key, signaling the reprise of the conflicts between Faust and Mephistopheles, or at the return of the introductory mysterious theme of the work. In the latter case, a third possible interpretation becomes legitimate: the *Sonata* appears as a great fresco, divided into two symmetrical halves, in which the first appears as an exploration that goes beyond tonality's Hercules pillars, and the second as a voyage back to Ithaca. A lengthily differed journey, just as Ulysses', that leads to a conciliatory and contemplative dimension. Liszt's *Sonata* thus appears to be in dynamic relation to the tradition from which it derives: it takes the complexity of formal correlation and structural tension from late Beethoven, but, in the renewed spirit of its time, it bends the conceptuality of these aspects in favor of a developing narrative approach. At last, with this masterpiece, Liszt traverses the *Sonata form* from the shore of apodictic judgments to the banks of narration, romance, and ultimately, epic.

(c) *Bach, Chopin, and narrative meaning*

Considering our last proposal of this interpretation—*remnant* as *renewal*—on a more abstract level, we can analyze the way in which Chopin applies the concept, already exploited by Bach with his *Well-Tempered Clavier*, of composing a cycle that includes all 24 major and minor keys. The type of sequence used by Bach, similar to the layout of an encyclopedia or a dictionary, with all terms alphabetically ordered, becomes radically reinvented a century later by Chopin in favor of a narrative order, much more fitting to the new Romantic sensibility.

The two Books of the *Well-Tempered Clavier*—the first of which written from 1718, the second around 1740—are a foundational work for tonal music, a *Magna Carta* for keyboard instruments. Both are comprised of 24 Preludes and Fugues, with each dyad being written in one of the 24 keys, according to this progression: n. 1 in C Major, n. 2 in C Minor, n. 3 in C# Major, n. 4 in C# Minor, n. 5 in D Major, and so on, climbing one-half tone at a time along the chromatic scale, with the major keys represented by odd and the minor ones by even numbers.

It seems appropriate to define this type of display as encyclopedic, given that there is no relation—in transitional cases from minor to major, i.e., from 2 to 3, or from 4 to 5—between the adjacent keys, if not that of being physically close on the keyboard. In addition, while the relevance between two consecutive encyclopedic entries is neither necessary nor excluded, the relation between two keys distant by only half a tone is always of the widest sort. Keys placed at one semitone from one another, although visually close on the keyboard, are in fact perceptually very far apart, as they are separated by no less than seven alterations—for precision's sake, by seven steps of ascending fifths. This is the reason why it is always complicated to modulate within the same piece in a convincing musical fashion to a key just a semitone away. The type of order chosen by Bach though responded quite perfectly to what the composer intended to achieve with his work, which was to bestow normative dignity to the newly invented equal temperament—the octave's subdivision into 12 identical semitones— and to illustrate, through the persuasive force of his great composition, the full potential of this new system.

After Bach, however, composers who attempted to create works that encapsulated the full palette of tonal harmony preferred to adopt a different criterion to organize the logical succession of every single key, searching for an "*internal harmonic connection between one key and the other, taking the circle of fifths as an outline.*" (Catucci 2012). The most emblematic case is that of Chopin and of his 24 *Preludes*, Op. 28. Setting aside all of the historical and aesthetic considerations that can be made on this extraordinary composition, what is important is the novelty of how the various tonalities follow one another and the perceptual influence that this new type of organizational layout has on the fruition of the work. Chopin's sequence is as follows: n. 1 in C Major, n. 2 in A Minor, n. 3 in G Major, n. 4 in E Minor, n. 5 in D Major, n. 6 in B Minor, and so on. The implemented principle thus dictates that the harmonic connection must proceed from a major key to its relative minor, with the following dyad placed an ascending fifth above. The technical consequence of this method is that the harmonic difference between one *Prelude* and the other is reduced to the absolute minimum.

There is no difference between a major key and its relative minor, as they are both composed of the same notes of a given scale, simply disposed in a different order. Keys which are placed a fifth apart, as C and G Major, or A and E Minor, have instead six out of seven shared notes. Thus, the transitional movement from one to the other is similar to that of a slight and steady rotation, like the slow passing of the celestial vault over our heads, through which the gradual key substitution takes place.

The consequence, from a phenomenological standpoint, is that the tonal kinship of the contiguous *Preludes* generates a narrative process, such that, even without any thematic connection between one piece and the next, there is a strong sense that each miniature resolves and flows into the following one, giving a great sense of continuity to the listening experience.

If Bach's encyclopedic order conveys a feeling of spatial extension, like a monumental palace, Chopin's system has a more transient quality, underlying the streaming of time and its narrative flow. Nonetheless,

> despite this structural difference, we can easily see an implicit homage to the Well-Tempered Clavier in Chopin's Preludes, an explicit claim of belonging to a tradition, a heritage and genealogy that has in Bach's oeuvre one of its most achievement. (Catucci, ibid.)

All of the musical masterpieces we have just discussed exhibit the rehabilitation of the past in a particular, essentially formal, perspective. The *renewal* of the *remnant* inevitably actuates a relation to history, articulated in different manners for every single case: with Schönberg, by unexpectedly reclaiming the adherence to tradition, with Ravel, just as paradoxically, by taking its distance from it, and with Schumann, Liszt, and Chopin by integrating it with a psychologically narrative and fluid principle, more fitting to Romantic sensibility.

13.3 Leftover as Unfinished

We would now like to explore the term *leftover* in a third acceptation, namely that of being *unfinished*. Toward that end, it is interesting to consider the case of the instrumental practice known as *figured bass*, which saw its most extensive use throughout the seventeenth and eighteenth centuries. *Figured bass* is the melodic line, often played on a harpsichord or an organ, placed in the bass clef, that serves as a foundation for the harmonies of a composition. The player was provided with only a simple written line and was expected to integrate and enrich it with improvised harmonies and cadences. Later on, composers started adding, in correspondence with the bass notes, a numeric series to the score, which indicated the type of chord that needed to be played, similar to the principle of jazz notation, which consists of a single melodic line and Roman numerals identifying specific chordal harmonies.

Why, though, did composers adopt such a laconic way of writing? Is it deliberately enigmatic, or is there another explanation? Certainly, this type of notation was easier to read and decipher, especially while improvising, and also to memorize. Moreover, the bass line already tacitly contains in itself all of the harmonic development, defined by the vocal or instrumental parts of the ensemble; for a musician accustomed to this aesthetic and this type of music, it becomes much easier to remain faithful to the expressive value of musical phrases by improvising on the essential melodic line, as opposed to deciphering a part with elaborately written details and secondary lines. The improvisatory nature of *figured bass*, made necessary by its synthetic writing, confers freshness and liveliness to every performance and anchors it to music's truer, deeper meaning, arguably much more so than a redundant score. The *unfinished* component in the notation of the *figured bass*, ultimately, is what ensures a specific form of liberty, namely that in which the performer is free to express not himself, but music's intrinsic truth, in the purest and most authentic manner.

We have thus far explored the various facets of music and its *residual* aspects. Returning to the jazz allusion made earlier, we would like to conclude this essay by, so to speak, closing the circle. As our discussion started with some reflections over the fact that the *remnant* can be interpreted in music as a form of *overabundance* of sense, it seems appropriate to implement this thought by showing how it can be applied to a tangible example of music.

Jazz pianist Brad Mehldau's 2018 album *After Bach*—a series of elaborations of J.S. Bach's Prelude and Fugues—seems a fitting example, for it contrapuntally implicates, as we shall see, many of the considerations made throughout this article. Mehldau's reinterpretations are a study in imaginative mastery and intellectual rumination: they are not the usual contemporary interpretations of Bach's profoundly fertile music, "jazzed up" *à la* Loussier, but rather a completely new chameleonic paraphrasing of extracts from the *Well-Tempered Clavier*. Mehldau does not riff on Bach's melodies, but, more interestingly, uses his methodology. By placing an unpaired Prelude or Fugue before his own composition, Mehldau explicitly creates a juxtaposition between an ancient and a modern approach to the practice of free improvisation. While both Bach's and Mehldau's compositions were certainly written down, we must not overlook how pivotal spontaneity and extemporaneity were in the *Kapellmeister's* musical life. Indeed, much of his work and fame grew around the complexity and virtuosity of his improvisations, and by recuperating this tradition, Mehldau deftly applies it to a frustratingly unknowable aspect of Bach's greatness.

The inevitable auditory comparison between the old and the new heightens and intensifies the meaning of the original eighteenth century compositions, by showing how much the latter owes to the previous, by exhibiting how deeply moving, poignant, elusive, and indeed, *ineffable* densely woven counterpoint can be, by ultimately revealing the ground shared by their composers, improvisers, and players. In short, by making implicit parallels explicit. Bach's inheritance of the apodictic insight as to how extemporary improvisation can enhance, rather than demolish, a systematic sense of musical architecture is picked up by Mehldau in a way that virtually attempts to delve into the emotional essence and contrails that the original compositions reveal to whomever carefully searches for them. Beauty truly is, in this case, in the eye of the beholder. The state of bliss conveyed by these partially extemporary and free improvisations allows every sound to be a teleologically unordered event, brimming with the power of an unexpected revelation, the striving puissance of incompleteness; and it is precisely this quality that reminds us of the way in which the *remnant* can be conceived as something *unfinished* and enables the performer to search for and express music's intrinsic truth.

If improvisation can certainly be a form of meditation on the moment, based on our past and present experiences, a sort of anchor to the *hic et nunc*, the improviser strives for none other than what, through him, becomes manifest to the listener's perception, which is the unfolding of music itself. In this sense, it is instructive to underline how the *finished/unfinished* dyad is etymologically equivalent to *perfection/imperfection*. As *perfect* means, in the purest sense, "something that is complete", without presupposing a comparison with anything else, its contrary, i.e., *imperfection*, does, therefore, not imply a value judgment, but simply a temporal quality, specifically associated with something that has not yet been completed and which is flowing at the moment.

Resuming our analysis of Mehldau's album, for instance, after a light and gentle performance of the 3rd Prelude in C# Major from Book I, he replies by proposing Bach's riff—originally in 3/8—in quite an unstable 5/8 rhythm and by conducting it into a harmonically audacious labyrinth, perhaps paying double homage to Bach by alluding to his *Kleines Harmonisches Labyrinth* (BWV 591). It is worth noting observe how the creation of a skewed metre, obtained by subtracting a note from the six original demiquavers, creates a drunken-like feeling, a sort of *imperfection* in relation to the original Prelude implemented by dissonant clashes, daring harmonies, and hallucinatory digressions. The core of this sensation is certainly the 5/4 m that seems *unfinished*, yet nonetheless, or better because of this, springs forth from Bach's intended time signature. As mentioned when discussing the *sonata form*, it is the dissolution of a structure, the subtraction of an element that adds meaning to a form, as a consequence of its potential possibilities. Similarly, a nostalgic and mysterious, rubato-heavy vision of the 12th Prelude and Fugue in F Minor from Book I is followed by a dreamlike meditation on some of the themes hinted at in Bach's original. Furthermore, after a solemn reading of the 16th Fugue in G Minor from Book II, Mehldau creates an awe-inspiring and embracing rhythmic feel, from which wobbling emotions, ghostly memories of the Fugue, swirl between the repeated and *ostinato* pedal notes, slowly and inevitably ending in silence. Huxley (1931) wrote that "*after silence, that which comes nearest to expressing the inexpressible is music.*": how blessed are we to experience them both here.

On the whole, though, the overall effect of all these revised readings is as if the pages from *The Well-Tempered Clavier* were turned upside down and seen through a broken and reassembled prismatic kaleidoscope. If Bach's Preludes and Fugues are like beautifully crafted logic puzzles, Mehldau's cryptic harmonies certainly appear instead as an insolubly deformed Rubik's Cube. Yet again, this is a value that adds not only to the new improvisation, but to the old inspiration itself. Defining Mehldau's attempt as a *Well-Tempered Jazz*- might perhaps be biting more off than we can chew, but what is certain is that these wonderfully crafted pieces have hopefully encompassed and epitomized our interpretation of the *remnant* as an extremely versatile concept. By being at the same time a *surplus*, a form of *renewal*, and of *imperfection*, we can surely claim that the beauty and mystery of music is not just tied to its acoustic seductiveness, but also to its conceptual, semantic, and philosophical implications. Indeed, as Beethoven said, "*music is a higher revelation than any other wisdom or philosophy*".

References

Boulez P (1966) Relevés d'apprenti. Èditions du Seuil
Catucci S (2012) The Accademia di Santa Cecilia in Rome concert leaflet
Damasio A (1999) The feeling of what happens.Body and emotion in the making of consciousness. Harcurt Brace
Huxley A (1931) Music at night and other essays. Chatto & Windus
Jankélévitch V (1961) La musique et l'Ineffable. Èditions du Seuil
Mendelssohn Bartholdy JLF (1878) Briefe aus den Jahren 1830 bis 1847. Mendelssohn
Ridley A (1995) Music, value and the passions. Cornell
Schönberg A (1911) Harmonielehre. Universal
Wittgenstein L (1953) Philosophische Untersuchungen. Suhrkamp

Chapter 14
For a Soft Law of Contemporary Project: Food for Thought and a Manifesto

Roberto Tognetti

> *When a system is far from equilibrium, small islands of coherence have the capacity to shift the entire system.*
> —Ilya Prigogine

Abstract This contribution starts with a survey on the reasons why the idea of planning is facing a crisis which is mainly linked to the effect of structural elements, such as the persistence of the hermeneutics of a 'long present'—which often becomes an alibi through nostalgic or inertial visions—and superstructural factors, such as the frenzy of regulations and procedures of our 'defensive bureaucracy'. As the pandemic is generating a new idea of future, partly out of enthusiasm and partly out of necessity, the concept of 'Soft law' is introduced as a new paradigm to turn the contemporary project into a device aiming at self-regulation, not new in itself but more and more necessary to stir things up in a field where everything is against planning. And the more so in a period when public debate is focussed on home planning, town planning and local planning again, and therefore, for a number of reasons, the achievement of outcomes cannot be avoided or concealed. Food for thought follows, besides a *Manifesto for responsible and creative Local Governance*, which lays the foundations for a disciplinary revision based on art. 118 of the Italian Constitution and aiming at applying the subsidiarity principle. The essay deals with some aspects of today's Italian situation but tries to face the problem at an international level.

Keywords Planning · Town planning · Local governance · Re-use · Urban regeneration

For a few decades, there has been a gap between reality and its interpretation in the local governance practice (and not only there). A cultural hegemony ruled by once dystopic and once diachronic components as a result of the persistence of methodological, operational and normative guidelines which date back to the past century as if they were still underlying the Fordist society and economy that generated

---

R. Tognetti (✉)
Fondazione Riusiamo l'Italia, Borgomanero, Italy
e-mail: r.tognetti@iperpiano.eu

L. Crespi (ed.), *Design of the Unfinished*, The Urban Book Series,
https://doi.org/10.1007/978-3-030-73457-2_14

them. Thanks to a number of factors, they have rooted in many fields connected with architecture and planning and even more so in management and regulatory practice, that is to say, its jumbled and confused regulations and procedures and the conformism of its managers and operators who are often on the defensive, thus generating the so-called 'defensive bureaucracy' approach.

The concept of 'long present' has crept into public debate at a certain point and it has only partly rooted in the founding purpose of "spreading ideas linked to the defence of memory, to the safeguard of intergenerational relationships and the creation of a world which is able to observe and face the complex problems of environment, economy and contemporary society in their wide geographical and temporal extension" (Brand 1999, 2009). This way it has become a new paradigm useful to justify a never-ending inertial condition, a relentless putting off of the status quo, a kind of pleased resigned acceptance of conservative policies, hardly ever separated from nostalgic shades or the acknowledgement of how serious the situation is (Benevolo 2012) Such a profile still shows up as an interpretation factor even in the vibrant renewal vision of the project theory when it is talked about "the viewpoints on the meaning of architecture after modernity and in the age of the long present when the foreseen town future never comes and what exists seems to be the last possible action context for architecture and towns (…). One of the main theoretical issues of modernity was the one about the best spatial synthesis between function and shape, today with IT revolution we face the opposite problem, i.e. to give meaning, story and uses—also temporary ones—to places that already have a shape. And in each situation, we have to turn the existent into a habitable, attractive and ecologically performant space, (…) This phase requires new paradigms (as new viewpoints on the future) and a new idea of planning physical space. It is a challenge that focusses on the value of the existent with conceptual devices dealing with the shifting of meaning and the life cycles of habitable spaces. A challenge that considers context as planning, landscape as an infrastructure that produces ecological value and the future of towns as a collective project and not as an authorial one" (Ricci 2019). And we find it again in terms of opposition (at last! we could say), when 'Policies for a different present' are sketched out in order to build a new idea of future by stating a future design that could be included in our "daily talk, in public debate and in our equipment as planners (*progettisti* in Italian) where the root word project shouldn't be seen as hot air or a mantra, but in its etymological meaning of the ones who throw themselves forward" (Carta 2019). The *Theory of the Black Swan* has come to reality (Taleb 2008) with the spreading of the Covid-19 pandemic, whose end cannot be seen yet, and it has contributed to reducing the asymmetry between reality and its interpretation, net of the denial movements and the increasing mistrust in science (McEwan 2020). In such dynamics, the idea of future has achieved central importance again both for those who share government responsibilities and for the myriads of citizens who now cannot avoid thoughts, ideas and conversations on how our lives will or may be in the next few years. We are back to talk about towns in the future, the relationship between villages and large cities or just the new exchange perspectives between town and country, of speeding up digital infrastructures, of sustainable mobility that can gradually free new public space, of new settlement models, of services to people in the new idea of a '15-min town', of a revised sanitary system with neighbourhood assistance, telemedicine and personalised treatment. We are back to talk of school and an educational and vocational system which will be enriched by blended learning and the multiplication of learning contexts, and much more.

This rebirth of the idea of future is important for us, as we have developed an original approach to temporary, creative and participated re-use (Campagnoli and Tognetti 2014) based on space rehab processes starting from demand, which are seen as action-research paths. Likewise, it is essential to speed up towards processes that push to a new paradigm in applying Laws and Regulations (Cacciari and Irti 2019). Turning to law is the missing element to legitimate a certain action or event. However, there is evidence that the less many human activities are regulated the more results they produce, especially in the field of creativity, experimentation and innovation. Thus, the definition of Nomos should be traced back to general guidelines and the prescriptive aspects should be attributed to different mechanisms according to a suitable balance between the various legal sources. Actually, the present reuse methods and techniques have often been carried out "at unchanged legal and normative arrangement". Any law or rule should act only through "general, open and non-prescriptive indications". With reference to reuse in the Italian system, the Consolidated Text on the subject of building has been recently modified with the addition of art. 23-quater about "Temporary Uses", whose first clause says: "The municipality can allow the temporary use of buildings and areas notwithstanding the current regulations in order to start processes of urban regeneration, of redevelopment of deteriorated urban areas, of recovery and development of buildings and abandoned or nearly abandoned urban spaces, while fostering the development of economic, social, cultural or environmental activities". Such elements as participation, sustainable, shared approaches, tension towards beauty and quality of the spaces and well-being of context and environment are inner rules in the conceiving and creating process among all the practice communities that have been working in the field of reuse for years: it is, therefore, the 'quality project' that dictates the rules. A new field where to introduce a Project and Local Governance Soft Law needs to be defined and its keywords and a manifesto will be introduced later on in this essay. According to Wikipedia, the term 'Soft law' refers to quasi-legal instruments which do not have any legally binding force. It should be used when we need to create a flexible discipline, which can adapt to the fast evolution of some sectors of economic or social life or incorporate within the national legal system Soft-law rules enforced by international organisations. (…) Such agreements do not create legal obligations between the parties (according to the principle of *pacta sunt servanda*) but only political commitment which is kept in compliance with the will of the parties. However, even actual acts can produce Soft-law rules in case they choose to impose not legally binding obligations to the recipient subject (soft obligation)".

The following statement on the value of synthesis in legal sources goes in the same direction: "Every Constitution can outlive the period in which it was generated, can fulfil the new protection needs, can regulate them through an evolutionary interpretation of its provisions. And this is a good thing because the *auctoritas* of the constitutional documents derives from their *venustas*, from the ability to go through the seasons of history. In order to manage to do so, an economy of signs of language is needed. And after all, this is the lesson we Italians were given by the Constituent Assembly in 1947: at least a quarter of the time was spent arguing to eliminate the unnecessary weight of the text, to reduce their weight. Just the opposite of what happens in Italy every day, when each budget law is written in half a million characters (Ainis 2020).

As it is shown by the most advanced experiences in the management of the processes of participation and development of new direct democracy forms, which coincidentally concern debate about general interest 'projects': "it is not a matter of shifting from public to private but of strengthening civil society by giving space and stamina to more articulate and less party-dominated forms of democracy. It is not a matter of 'reforming' bureaucracy, it is a matter of getting rid of it" (Sclav 2020).

Thus, the Soft-law open concept can be suggested as a general principle which should be aimed at establishing a new way of regulating the town and region transformation processes. This will not include the simplistic seductions of the so-called 'Genua Model' (Lanza 2020). And if such enthusiasm helps us to define the conditions of the 'how', on the other hand, the 'when' issue becomes urgent and unavoidable, both if it is considered in terms of digital transition: "We are coming to terms with Game, with digital civilisation (…). In this sense, Covid-19 looks like the rehearsal for the next game level, the final mission: saving the planet" (Baricco 2020) and if it is considered from the point of view of climate change: "Let's acknowledge that we cannot stop the climate catastrophe any longer. Climate Apocalypse is coming. In order to face it, we need to acknowledge that we cannot prevent it" (Franzen 2020).

In the period of 'different present', the culture of architectural, environmental, landscape or local project must find its place in the public life of the country. The discipline needs to be brought back to its civil foundations of service to people, communities, towns and regions. The old-fashioned tool box needs to be updated: zoning, costs for the town, quantitative standard, duration of the adoption/approval process, the idea itself of 'mode of use' in its strictest meaning are all parameters that are less and less compatible with the increasing complexity of the problems to be faced to ensure simplicity, participation, sustainability, flexibility, generativity of Plans and Projects. Among the most striking features of this renovated general context, we could highlight:

A. The fall in population and the shrinking of settlements, which generate large areas of increasing porosity, where voids (abandoned, underused, vacant, unsold, unfinished spaces) outnumber full spaces and produce endemic 'shrinking' phenomena, which is more or less the opposite of the building boom that took place from the 1960s to the first years of the new millennium.
B. The shifting towards a new socio-economic model which hasn't come to an end and has brought about a never-ending condition of slow and often incomprehensible transition where most of the Plan predictive features do not stand up anymore, where the future is often neither predictable nor questionable and

where large portions of society and economy float in uncertainty and insecurity, someone as a choice and someone, more frequently, out of need.

C. The gradual acceleration of digital paradigms, which means town planning is given a conciliation role between the new and old world, i.e. between the ones who master their use and those who are unskilled, or the ones who can only make a partial limited use of it. Augmented reality, Internet of things, big data, blockchain, use of drones, machine learning, robotics are only some of the changes that are taking shape. Many of them could have revolutionary functions in various aspects linked or linkable to a new way of planning towns.

D. Social impact has been a guiding principle for a number of policies for long. From infrastructures to investments in key sectors, the measuring of social impact is considered as a diriment. Such factor in the form of ESG (Environmental, Social and Governance), Social Balance, Public–Private Partnership and much more is increasingly involving a lot of subjects in profit enterprises in the searching of new models of wealth redistribution and of welfare improvement. Measuring the impact is also one of the most reliable techniques to analyse the susceptibility of an investment in terms of fulfilment of general interests. The more reliable you are in ensuring the prefixed impact, the more able you are to guarantee the investment effectiveness and usefulness to the possible public, private or mixed financing subjects.

Therefore, architecture and town planning need to be free again, planners need to feel pleasure in their role, as well as politicians and public managers, who play an even harder role in the process. The regulatory frenzy that has turned town planning into a formal system distant from a 'project of town' should be strongly opposed. Planning needs to be people-centred again, by means of simple words such as house, street, square, workshop, factory, neighbourhood, town. "What is the City but the People?", Shakespeare said, but what should we say while looking at the empty urban spaces? Its overturning leads to a question: aren't buildings without people a non-town?

We conclude with a *Manifesto for a responsible and creative Local Governance*, which lays the foundations for a disciplinary revision based on art. 118 of the Italian Constitution and aiming at applying the subsidiarity principle.

14.1 Manifesto for a Responsible and Creative Local Governance

This Manifesto was conceived by iperPIANO (www.iperpiano.eu) in April 2015, as a follow-up of the research work used to write the book *Riusiamo l'Italia*, quoted above. Its text follows.

With reference to the Siena Declaration in 1309, "among the studies and cares that town governors should have, of paramount importance is the beauty of the city, because the city must be suitably decorated in order to give pleasure and happiness to strangers, and for the honour, prosperity and improvement of the town itself and its citizens".

Hoping for the achievement of a culture of responsibility based on cooperation between insiders, citizens and town management and inspired by the following values and guidelines (www.labsus.org):

a. Mutual trust: Despite the public prerogatives in the fields of supervision, planning and control, town management, insiders and active citizens relate to one another in terms of mutual trust, assuming that their cooperation aims at general interest achievements.
b. Publicity and transparency: the town management ensures to publicise the cooperation opportunities, the suggestions received in the planning phases, the allocated forms of support, the decisions made, the achievements and evaluations. Transparency is recognised as the main method to ensure impartial behaviours towards insiders and active citizens and verifiability of the actions taken and the outcomes achieved.
c. Responsibility: responsibility is the recognised focus in the relationship between town management, insiders and citizens, and it is the necessary prerequisite for outcome-oriented cooperation.
d. Proportionality: the town management correlates the administrative acts, the guarantee and the quality standards required for each authorisation procedure to the actual need to protect public interests. In so doing, it defines cases and conditions so that the parts involved are allowed to use the simplest procedures.
e. Adequacy and differentiation: the cooperation forms between insiders, citizens and town management are suitable to the context of sustainable development needs.
f. Informality: the town management requires formal procedures in the relationship with insiders and citizens only if provided for by law. In all other cases, it ensures flexibility and ease in the relation, provided public ethics is respected, as per the civil servants' code of practice and the principles of impartiality, good practice, transparency and certainty.
g. Civil autonomy: the town management recognises autonomous initiative to insiders and citizens alike and takes all the necessary measures to make sure it can be employed by active citizens and their organisations.

Having recalled the international treaties in the fields of sustainability, landscape, town and historical centres, protection and promotion of heritage and environment, acknowledgment of cultural values by the community;

Having emphasised the need for a national Town Planning Reform Bill (Campos Venuti 2013), the following is stated.

14.2 Manifesto for a Responsible and Creative Local Governance

On its own or in association with others, every municipality is autonomous in defining the best administrative procedures that guarantee:

1. The quality, decency and style of every kind of buildings, landscape and urban setting;
2. Housing dimensional standards and their layout, in terms of functions, proportions, adaptability and use;
3. The hygienic conditions of light, air, comfort and well-being through the best available technologies, including the retrieval of natural techniques;
4. The performance standards in terms of eco-sustainability and energy efficiency in buildings, to be ensured thanks to the best available technologies, including the retrieval of natural techniques, and tailored for specific restoration works of historical and artistic heritage;
5. The implementation of criteria of planimetric/volumetric regulations thanks to planning indexes based on homogeneous elementary parameters, which include a system of rounding to the nearest figure and/or an automatic adjustment of minor increases, or, if applicable, an automatic adjustment by progressivity;
6. The implementation of town planning standards based on non-quantitative criteria, which should be defined according to multicriteria parameters consistent with the town quality and the life quality requested by the inhabitants, including the evaluation of their impact;
7. The implementation of specific taxes based on elementary calculation criteria, i.e. the smallest possible number of parameters, so that they are easily understood, transparent and verifiable;
8. The turning of the building licence into a credit note that can be used in flexible, progressive and proportional ways, both in quantitative terms to ensure the maximum possible equity and as a bonus correlated to features of innovation, experimentation, complexity and beauty of the works.
9. Paths of development, adoption and approval of public interest plans, programmes and projects that provide for widespread ways of informing and involving citizens and stakeholders;
10. The use of management methodology based on information and telematic systems, such as the local information systems (SIT) or the Geographic Information System (GIS), the data gathering and processing (OpenData) to be used as open source and in the easiest possible way for both insiders and general public;
11. The management of the application, monitoring and evaluation process, including supervision and sanctions, based on the above-mentioned SIT, GIS and OpenData systems, besides further technologically advanced instruments, such as satellite photography and the use of suitably equipped drones to measure and detect the building plans and volumes.

12. Measures to fight land consumption, i.e. guidelines to limited, selective and justified consumption which should be correlated to the renovation of abandoned areas or underused spaces, which is to be privileged. In such perspective, ways of progressive adjustment of the selective densification of the existent could be experimented, according to the principle of "the more you use, the less you pay".
13. Superior institutions such as State and regions monitor, evaluate, support, foster and promote innovation and experimentation processes thanks to the exchange, publicity and transferability of good practice. To this purpose, each of them appoints a specific mission structure which is supported by a forward planning unit that can be subject to phases of attuning and complementarity;
14. The monitoring and evaluation outcomes allow to define efficiency levels for local policies, which may lead superior institutions to decide on substitution or external administration;
15. The fight against the most severe and rooted forms of unlawful building must involve the power of quick substitution, because of the personal liability public administrators may come up against, up to the deployment of the army to carry out demolition and restoration to the original condition.

References

Ainis M (2020) Il diritto a Internet che già c'è. La Repubblica

Baricco A (2020) Virus: è arrivato il momento dell'audacia. La Repubblica

Benevolo L (2012) Il tracollo dell'urbanistica italiana. Laterza, Roma

Brand S (1999, 2009) Il lungo presente: tempo e responsabilità. Mattioli 1885, Fidenza

Cacciari M, Irti N (2019) Elogio del diritto. La Nave di Teseo, Milano

Campagnoli G, Tognetti R (2014) Riusiamo l'Italia. Da spazi vuoti a start-up culturali e sociali. Gruppo 24 Ore, Milano

Campos Venuti G (2013) Polemica disciplinare e assenze politiche. Urbanistica Informazioni, Roma

Carta M (2019) Futuro. Politiche per un diverso presente. Rubbettino, Soveria Mannelli

Franzen J (2020) E se smettessimo di fingere?. Einaudi, Torino

Lanza V (2020) Come funziona il modello Genova per sbloccare i cantieri. Il Riformista. Available https://www.ilriformista.it/come-funziona-il-modello-genova-per-sbloccare-i-cantieri-90505/

McEwan I (2020) Invito alla meraviglia. Einaudi, Torino

Ricci M (2019) Habitat 5.0. L'architettura del Lungo Presente. Skira, Milano

Sclav iM (2020) Al di là del muro della burocrazia. https://sbilanciamoci.info/al-di-la-del-muro-della-burocrazia/

Taleb NN (2008) Il cigno nero. Il Saggiatore, Milano

Printed by Printforce, the Netherlands